Dams in Japan

Dams in Japan

Past, Present and Future

Japan Commission on Large Dams

CRC Press
Taylor & Francis Group
Boca Raton London New York Leiden

CRC Press is an imprint of the
Taylor & Francis Group, an **informa** business

A BALKEMA BOOK

The preparation of this text was partly subsidized by the River Fund of the Foundation of the River & Watershed Environment Management, Japan.

Cover illustrations credits:

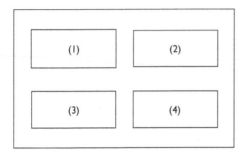

(1) Tokuyama Dam (ER, 161 m, #75), by courtesy of JWA

(2) Honen-ike Dam (CB, 30 m, #110), photo taken by Mr. Ishikado Naoyosi (JDEC Dam Photo Contest, 2005)

(3) Miyagase Dam (PG, 156 m, #52), by courtesy of MLIT

(4) Kurobe Dam (VA, 186 m, #58), by courtesy of KEPCO

First issued in paperback 2017

Taylor & Francis is an imprint of the Taylor & Francis Group, an informa business

© 2009 Taylor & Francis Group, London, UK

Typeset by Vikatan Publishing Solutions (P) Ltd, Chennai, India.

Published by: CRC Press/Balkema
              P.O. Box 447, 2300 AK Leiden, The Netherlands
              e-mail: Pub.NL@taylorandfrancis.com
              www.crcpress.com – www.taylorandfrancis.co.uk – www.balkema.nl

Library of Congress Cataloging-in-Publication Data

Dams in Japan : past, present, and future/Japan Commission on Large Dams.
    p. cm.
  Includes bibliographical references and index.
  ISBN 978-0-415-49432-8 (hardcover : alk. paper)

1. Dams -- Japan -- Design and construction -- History. 2. Dams -- Environmental aspects -- Japan -- History. 3. Dams -- Social aspects -- Japan -- History. I. Nihon Dai Damu Kaigi. II. Title.

TC558.J3D32 2009
627'.80952--dc22   2008051275

ISBN 13: 978-1-138-11454-8 (pbk)
ISBN 13: 978-0-415-49432-8 (hbk)

Contents

List of Tables

List of Figures

Preface

The geographical features of Japan are quite distinct, especially when compared with those of continental countries. Collision of four tectonic plates brought about the narrow archipelago of Japan, its high orogenic and volcanic activities, complicated geological structure, and frequent occurrence of large-scale earthquakes. The existence of backbone mountain ranges formed many small river basins. Plentiful precipitation characteristic of the Asian Monsoon climate fostered rich forest zones with clear streams on mountain sides. It is remarkable that two thirds of the national land of Japan, one of the most industrialized counties in the world, is still covered by green forests. This precipitation also resulted in the development of alluvial plains along the downstream parts of rivers through repeated flooding. These plains are now the major areas of social and economic activities for the Japanese.

The high productivity of paddy cultivation, which suits natural conditions in Japan, supported Japanese society as its key industry for more than ten centuries until the achievement of industrialization. Its experience and innovations in water management and flood disaster mitigation including construction and management of irrigation ponds became an important part of the social framework of Japan.

The modernization of Japan started in the latter half of 19th century, and was successfully achieved within a short period of 100 years or so. This progress was accompanied by the urbanization and industrialization of the alluvial plains. One infrastructure development important for this success was protection from flood disasters by large-scale river improvement works and dam projects, and the supply of necessary water and energy by dams.

The society and economy of Japan are now mature, following a period of rapid economic growth. The general public now holds a harsh view of major social infrastructure such as dams. The population of Japan dramatically increased from 40 million at the end of 19th century to 128 million at the beginning of 21st century, and the Japanese are now facing a new phase: population decline.

Dam construction projects in Japan peaked about ten years ago. The effective use and redevelopment of existing dams and prolonging their operational life are major tasks for the coming years. In addition, measures to adapt to climate change accompanying the global warming are emerging as world-scale challenges. In the face of fear of more frequent floods and droughts, the role of dam reservoirs in mitigating the fluctuation of the hydrological cycle will be of greater importance in the future.

At this turning point for Japanese society and its economy, the Japan Commission on Large Dams (JCOLD) has attempted to summarize the roles of dams in Japan.

The sub-committee, formed in 2003, finalized the text in Japanese after three years' effort by its members in 2005. This work covers their long history starting from 7th century when irrigation ponds were first constructed for paddy cropping, until the beginning of the 21st century. It also elaborates the entire range of the roles of dams: water supply, power generation and flood control. Moreover, the work tries to clarify the negative impacts of dams on the natural environment and local societies, as well as extensive efforts made to minimize these impacts. Subsequently in 2006, "The Story of Dams in Japan" was published as an overview based on this work.

This book in English was prepared based on the original version in Japanese in order to transmit the Japanese experience to people concerned with water resources management and dam projects around the world. One year's activity of the working group organized by JCOLD resulted in this book which contains much information not well-known to foreigners. We are genuinely hopeful that this book is utilized widely abroad.

Yoshikoshi, Hiroshi
President, Japan Commission on Large Dams
(Adviser, Tokyo Electric Power Company Inc.)
March 2009

The preparation of the English version text

Although this book is based on the original Japanese language text, it is not a direct translation but the result of deliberate examination and amendment performed as follows.

First, the context and wording are reviewed and organized for the better understanding of foreign readers who are not fully aware of the Japanese background. The parts deemed to be too much detailed are summarized or abbreviated. Elementary technical explanations were also abbreviated, since the readers are supposed be involved in water resources management and dam projects.

Then, a very brief history of Japan was added to the exordium along with four referential materials, including an outline of the administrative system of Japan and the basic framework of river management, which were prepared at the end of the book. These are intended to provide basic information for readers.

In addition, the specifications of dams which appear in this book are listed in a table, and a location map was drawn.

This English version publication is an outcome of intensive activities of the members of the working group with the assistance of the Japan Translation Center Ltd.

Hamaguchi, Tatsuo
Chairman, Working Group for the Preparation of the
English Version Text of "Dams in Japan",
JCOLD (2007–2009)
(President, Japan Dam Engineering Center)
March 2009

Members of the Working Group for the Preparation
of the English Version Text of "Dams in Japan", JCOLD (2007–2009)

Name	Affiliation	
Hamaguchi, Tatsuo	Japan Dam Engineering Center	Chairman
Okano, Masahisa	Japan Construction Training Center	Vice Chairman
Mori, Hideto	CTI Engineering Co., Ltd.	Chief Secretary
Ozaki, Yoshifumi	CTI Engineering Co., Ltd.	Secretary
Fukuroi, Hajime	KEPCO[1]	Previous Member
Hori, Yoshiro	River Bureau, MLIT[2]	Previous Member
Horiguchi, Kazuhiro	Nuclear and Industrial Safety Agency, METI[3]	Member
Isayama, Ai	Water Resources Engineering Dept., JWA[4]	Member
Ito, Hiroshi	IDEA Consultants Inc.	Member
Jojima, Seishi	Member of the Sub-Committee	Member
Kano, Shigeki	Dam Dept., JWA	Previous Member
Kondo, Masafumi	Japan Dam Engineering Center	Previous Member
Mitsuzumi, Akira	KEPCO	Member
Miyagawa, Kenji	Rural Development Bureau, MAFF[5]	Previous Member
Mizuhashi, Yutaro	J-POWER	Member
Nishino, Noriyasu	Rural Development Bureau, MAFF	Member
Nishio, Toshiya	JIID[6]	Member
Okamoto, Kazuyoshi	River Bureau, MLIT	Member
Okuda, Akihisa	River Bureau, MLIT	Member
Shimamoto, Kazuhito	Japan Dam Engineering Center	Member
Takahashi, Akira	TEPCO[7]	Member
Tsuchiyama, Shigeki	Chubu Electric Power Company Inc.	Previous Member
Yamamoto, Keiichi	River Bureau, MLIT	Previous Member
Yamawaki, Tsukasa	Chubu Electric Power Company Inc.	Member
Fukuoka, Shoji	Professor, Chuo University	Advisor
Ishikawa, Tadaharu	Professor, Tokyo Institute of Technology	Advisor
Matsumoto, Norihisa	Japan Dam Engineering Center	Advisor
Tanaka, Tadatsugu	Professor, University of Tokyo	Advisor

Remarks: The listed affiliations are those at the time of working group activity.

[1] Kansai Electric Power Company Inc.
[2] Ministry of Land, Infrastructure, Transport and Tourism
[3] Ministry of Economy, Trade and Industry
[4] Japan Water Agency
[5] Ministry of Agriculture, Forestry and Fisheries
[6] Japanese Institute of Irrigation and Drainage
[7] Tokyo Electric Power Company Inc.

Preface of the original text in Japanese: The preparation of this work

Human life has been closely linked to rivers in various ways, but our development of dam technology that lets us store river water has, by enabling us to obtain a larger and more stable supply of water and to effectively control floods, been a much more revolutionary event in the history of this relationship than our discovery of ways of simply drawing water from rivers or protecting our land from flooding by constructing levees. In this way, dams have played major roles in contributing to human progress and in achieving progress in water usage and flood control at critical junctures in the history of human society.

In recent years, opinions and media reports that view dams skeptically have become conspicuous: concern with the cost and benefit of dams, their impact on social and natural environments and pressure on the finances of national and local governments.

Nevertheless, destruction caused due to large-scale floods or droughts occurs repeatedly in various parts of Japan almost every year.

But the roles that dams have borne in response to the demands of each age are now taken for granted, and it appears that the desires of the river basin residents who pioneered each new era and the enthusiasm for dam construction of the people of each age are being forgotten as events in the distant past. We have to inform future generations of these historical facts.

It has also been impossible to ignore the impacts of the construction and existence of dams on their surrounding regions and environments, but we believe that the public is not adequately aware of efforts which were made by those who planned, constructed and managed dams to deal with these impacts. We wish to inform the public of these historical efforts.

The Japan Commission on Large Dams (JCOLD) recently responded to these circumstances by establishing the Subcommittee on the Survey of the Roles of Dams under its Technology Committee in 2003, and this new subcommittee has begun surveys to again clarify the roles of dams. In the midst of sweeping changes as we enter the twenty-first century, dam construction and management projects are also facing a major turning-point. At this time, it is extremely important that we look back at the roles of dams constructed over the past 2000 years and strive to reorganize the concepts of dams and water problems so that water usage can be sustained in the future.

This survey was conducted by volunteers connected with the JCOLD that constructs and manages dams, and also involved reviewing reports of similar surveys and documents in the past. Thanks to the efforts of the committee members and the cooperation of concerned bodies, this work has been completed.

I am hopeful that as many as possible of the young people who will lead the next generation and those who guide them will read this work. Nothing would delight me more than hearing the views of readers and stimulating debate among them about its contents.

Okano, Masahisa
Chairman, Subcommittee on the Survey of the Roles of Dams
Japan Commission on Large Dams (2003–2005)
(Vice President, Japan Construction Training Center)
March 2009

Members of the Subcommittee on the Survey of the Roles of Dams, JCOLD (2003–2005)

Name	Affiliation	
Okano, Masahisa	WEC[1]	Chairman
Nanami, Yoshiaki	WEC	Previous Secretary
Omoto, Iemasa	WEC	Secretary
Urakami, Masato	WEC	Previous Secretary
Hiroki, Kenzo	River Bureau, MLIT[2]	Member
Hongo, Naofumi	JIID[3]	Previous Member
Jojima, Seishi	Member of International Committee, JCOLD	Member
Kakizaki, Tsunemi	River Bureau, MLIT	Member
Katayama, Mitsuya	Dam Dept., JWA[4]	Member
Kawanaka, Masamitsu	Rural Development Bureau, MAFF[5]	Member
Kikuchi, Koichiro	J-POWER	Member
Kitamura, Tadashi	Operation and Maintenance Dept., JWA	Member
Kobayashi, Junji	TEPCO[6]	Member
Matsuda, Fumihide	Rural Development Bureau, MAFF	Previous Member
Noike, Etsuo	Chubu Electric Power Company Inc.	Member
Nukina, Koji	Japan Dam Engineering Center	Member
Numata, Hiroo	Nuclear and Industrial Safety Agency, METI[7]	Member
Okamura, Yukihiro	River Bureau, MLIT	Member
Seto, Taro	JIID	Member
Takamura, Yuhei	River Bureau, MLIT	Previous Member
Tanida, Hiroki	River Bureau, MLIT	Previous Member
Terazono, Katsuji	CTI Engineering Co., Ltd.	Member
Yonezaki, Fumio	Dam Dept., JWA	Previous Member
Yoshida, Nobuo	Operation and Maintenance Dept., JWA	Previous Member
Yoshizawa, Kazumi	Nuclear and Industrial Safety Agency, METI	Previous Member

Remarks: The listed affiliations are those at the time of subcommittee activity.

[1] Water Resources Environment Technology Center
[2] Ministry of Land, Infrastructure, Transport and Tourism
[3] Japanese Institute of Irrigation and Drainage
[4] Japan Water Agency
[5] Ministry of Agriculture, Forestry and Fisheries
[6] Tokyo Electric Power Company Inc.
[7] Ministry of Economy, Trade and Industry

Abbreviations

ORGANIZATION

CEPCO	Chubu Electric Power Company, Incorporated
ENERGIA	The Chugoku Electric Power Company, Incorporated
FEPC	The Federation of Electric Power Companies of Japan
HEPCO	Hokkaido Electric Power Company, Incorporated
Hokuriku EPCO	Hokuriku Electric Power Company, Incorporated
J-POWER	Electric Power Development Company, Ltd
JWA	Japan Water Agency
KEPCO	The Kansai Electric Power Company, Incorporated
Kyushu EPCO	Kyushu Electric Power Company, Incorporated
MAFF	Ministry of Agriculture, Forestry and Fisheries
METI	Ministry of Economy, Trade and Industry
MLIT	Ministry of Land, Infrastructure, Transport and Tourism
MOC	Ministry of Construction
MOE	Ministry of the Environment
OEPC	The Okinawa Electric Power Company, Incorporated
Pref.Gov.	Prefectural Government
TEPCO	The Tokyo Electric Power Company, Incorporated
Tohoku EPCO	Tohoku Electric Power Company, Incorporated
YONDEN	Shikoku Electric Power Company, Incorporated
WRDPC	Water Resources Development Public Corporation

TYPE OF DAM

AFRD	Asphalt Faced Rockfill
CB	Buttress
CFRD	Concrete Faced Rockfill
ER	Rockfill
HG	Hollow Gravity
PG	Gravity
TE	Earthfill
VA	Arch

PURPOSE

C	Flood control
H	Hydroelectric
I	Irrigation
N	Normal functions of the river water
S	Water supply

Remarks: When a dam's name is followed by bracketed information, e.g. "Sakuma Dam (PG, 156 m, #82)", the three items in brackets are dam type, dam height, and dam number in the table "the Main Dimensions of Dams described herein".

Chapter 1

Introduction

1.1 CIRCUMSTANCES SURROUNDING DAMS[1] AND THE AIMS OF THIS WORK

Japan is the first nation in the Asian Monsoon zone to achieve modern industrialization and it did this in the relatively short period of about 100 years, beginning in the late nineteenth century. The reasons for this include several factors related to Japan's history, its geography, and the character of its people. From the perspective of development and economics, the synergistic effects of concentrated investment in transportation, communication, energy, disaster prevention and hygiene infrastructures accompanied by the stimulation of private sector economic activities, played a major role. From the perspective of the development and utilization of the national land, a major factor behind Japan's success was utilizing the country's alluvial plains that are essentially vulnerable to flood disasters as national land to promote industrialization and urbanization, making national efforts toward controlling rivers to prevent flood and drought disasters as well as to utilize their water resources.

Dams played big roles in this process. Factors contributing to this success include a long pre-modernization history of water usage, primarily for paddy field agriculture. Dams were constructed and operated to form many irrigation ponds, beginning as early as the seventh century.

Dams supported social and economic development that preceded modernization and have been viewed as symbols of modernization and of social vitality that utilizes nature. This is true not only of Japan, but also of other countries, and throughout the world, dams were constructed without objection.

However, in recent years, perspectives on dams have changed substantially. Since the 1980s, dam projects have attracted criticism from the public, because their substantial impacts on society and the natural environment are becoming clear. Through

---

1 Dams:
  a In its narrow meaning, a structure shaped like a levee that blocks the flow of a river to raise its water level to store water. In its broader sense it includes the reservoir that is formed by a dam. In Japan's River Law, special regulations are applied to dams with a height of 15 m or more from the foundation ground to dam crest.
  b The International Commission on Large Dams (ICOLD) defines a dam with height of 15 m or more from the foundation ground to dam crest as a large dam.

Presented by Osaka Prefectural Government

Figure 1.1.1 Sayama-ike Dam (TE, 18.5 m, #A): The oldest dam in Japan still in operation. (See colour plate section)

numerous media reports, books and large quantities of information on the Internet dealing with the impacts of dams on people, river basins and ecosystems or the economic effectiveness of dams, questions about dams have escalated and this has expanded into a debate on the overall cost-benefits of dam projects.

Many regions of the world continue to suffer from a shortage of water resources. There are limits to rainfall and river flow rates that are usable as water resources around the world. And because there are also temporal and regional differences and fluctuations, dam reservoirs shall be managed appropriately. Even in Japan, floods and droughts in scattered locations cause heavy losses almost every year. We must fulfill our responsibilities to our descendants by utilizing water resources that have already been developed by dams in the past to achieve sustainable water usage that is balanced with environmental conservation and in this way, prevent harm to future generations.

This work considers this situation by first outlining the state of dams in Japan incorporating comparisons with dams around the world. Next, it looks back at the roles of dams that have been constructed and operated throughout human history, centered on the history of water usage as it pertains to the limitations of dams. This is followed by an account of the negative impacts of dams and clarification of the efforts made in Japan during the past half century to mitigate these negative impacts. The work builds on this by reorganizing concepts of future water problems and dams, including an international perspective, to portray the roles of dams in the twenty-first century.

We entreat people conducting dam projects and managing dams, or people who are concerned with these activities, to read this work and consider its contents in relation to dams with which they have been directly involved.

1.2   A SHORT HISTORY OF JAPAN [1, 2]

The Japanese archipelago is located at the eastern end of the Eurasian continent and surrounded by the ocean. Its highest mountain is Mt. Fuji (elevation 3,776 m) and it experiences four distinct seasons. Japan has a long history of habitation by people who have skillfully utilized and transformed the land, water and nature to expand its production and living space. The historical background to Japan's development through dependency on dams is discussed in each chapter. This section is an outline of Japanese history that focuses on social systems, economics and the living infrastructure during each age of dam construction. Table 1.2.1 is a chronology of Japanese history.

Table 1.2.1 Chronology of Japanese history [1].

Western calendar	Major periods of Japanese history
	Paleolithic (pre-10,000 B.C.)
10,000 B.C.	
	Jomon (ca 10,000–ca 300 B.C.)
300 B.C.	Yayoi (ca 300 B.C.–ca A.D. 300)
A.D. 300	Kofun (ca 300–710)
	Asuka (593–710)
400	
500	
600	
700	
	Nara (710–794)
800	
900	
	Heian (794–1185)
1000	
	Fujiwara (894–1185)
1100	
1200	
	Kamakura (1185–1333)
1300	
	Muromachi (1333–1568)
1400	
	Northern and Southern Courts (1337–1392)
1500	
	Sengoku (1467–1568)
1600	Azuchi-Momoyama (1568–1600)
1700	
	Edo (1600–1868)
1800	
1900	Meiji (1868–1912)
	Taisho (1912–1926)
	Showa (1926–1989)
	Heisei (1989–

1.2.1 The Jomon and Yayoi Periods
(to approximately A.D. 300)

In Japan, there are traces of a Paleolithic age: human habitation earlier than 30,000 B.C. The period when Jomon pottery (pottery decorated with patterns of knots) was used is called the Jomon Period (approx. 10,000 B.C. to 300 B.C.). There are traces of evidence that during this period, people hunted and gathered food, and formed large villages where they cultivated chestnuts and similar foods.

The Jomon Period was followed by the Yayoi Period when Yayoi Pottery was used (approx. 300 B.C. to A.D. 300). During this period, rice cultivation was transmitted from the Korean Peninsula to northern Kyushu by migrants arriving from the continent and was then transmitted to western Japan. At about the same time, metal and iron implements came into use. According to the Chinese historical record, Record of the Three Kingdoms, Book of Wei (approx. A.D. 280), Japan was in a state of conflict between more than 100 provinces, but later it was unified under the name Yamatai (approx. A.D. 180). However, the details of this process are still not clear.

1.2.2 The Kofun – Nara Period
(approx. A.D. 300 to 794)

This period when *kofun* (massive burial mounds made of embanked soil) were constructed, primarily in the Kansai Region, is called the Kofun Period (approx. A.D. 300 to 710). The construction of these massive *kofun* indicates that the stratification of agricultural society had progressed since the Yayoi Period.

During this period, Buddhism and the Chinese writing system were imported from the continent, the Yamato Court was established, uniting the entire country, and a government based on *ritsuryo*, a set of Chinese style laws, ruled. At the end of the Kofun Period, the capital was located in Asuka in Nara, so this stage is called the Asuka Period (A.D. 593 to 710).

Prince Shotoku (574 to 622) served as regent to the Empress Suiko, achieved the court rank, 12th grade Cap Rank, and promulgated the Seventeen Article Constitution (604). A convert to Buddhism, he worked to expand the religion and founded Horyuji Temple (607) that has survived till the present day.

After the death of Prince Shotoku, Prince Naka-no-Oe (626–672) took the throne as the Emperor Tenji, enacted the Four Edicts concerning land control and the structure of government and promoted the concentration of power in the central government. This is called the Taika Reform.

After Emperor Tenji abdicated the Imperial Throne (672), his younger brother succeeded under the name Emperor Tenmu (672 to 686). He continued to reform the recruitment of officials for the central and regional governments. In 689, he began the compilation of the Asuka Kiyomihara Code that became the foundation for the all-inclusive Ritsuryo system (Japanese legal system).

The period when the large capital, Heijokyo (4.3 × 4.8 km), was located in Nara, is called the Nara Period (A.D. 710 to 794). The Ritsuryo government matured and Buddhism was given formal status as the state's religion. Official provincial monasteries (*Kokubunji*) and nunneries (*Kokubunniji*) were established throughout Japan to display the authority of centralized government, and the Todaiji Temple that housed Japan's largest statue of Buddha was established in Nara. The culture established during this period is called the Tenpyo Culture, which produced Japan's oldest historical chronicles, the Kojiki (Record of Ancient Matters) (A.D. 712) and Nihonshoki (A.D. 720) (Chronicle of Japan), and Japan's oldest collection of poetry, the Man-yoshu (A.D. 759).

During this age, agricultural irrigation ponds were widely constructed. The Sayama-ike Pond (#A) is the oldest recorded irrigation pond (A.D. 616). After numerous reconstructions, this pond is still in use.

The Konden-Einen-Shizai Law (A.D. 743) that confirms ownership of land by its developer was enacted, spurring the creation of agricultural land. This law also encouraged the development of surveying techniques, contributing to the formation of the land at Heijokyo.

1.2.3   The Heian Period (794 to 1185)

During the Heian Period, the capital was in Kyoto (Heiankyo) and Kyoto remained the capital of Japan from then until 1868.

The Fujiwara family who were closely linked to the Imperial Court monopolized administrative power, building a golden age by creating an indigenous aristocratic culture, based on Chinese culture. The period from 894 to 1185 is, therefore, called the Fujiwara Period. But as their political authority weakened, discontented farmers formed links with armed groups of warriors known as "*bushi*".

The Taira and Minamoto families who led two powerful armed groups clashed repeatedly, but the Taira family finally emerged victorious, grasping authority under the court and prospering for twenty years.

The defeated Minamoto leaders mustered their strength in the Tohoku Region of Japan and attacked the capital again, finally destroying the Taira Family.

During this period, regional development was generally slow, but monasteries that maintained contact with Chang'an in China introduced external culture, contributing greatly to Buddhist activities and the provision of infrastructure. The monks Kukai and Gyoki are particularly famous. Kukai, who positioned helping the poor as a part of Buddhist religious practice, worked hard to develop agricultural land and irrigation ponds.

1.2.4   The Kamakura to the Azuchi-Momoyama
Period (1185 to 1600)

The chieftain (Minamoto no Yoritomo) of the Minamoto Family who was victorious in the Minamoto – Taira war, obtained the title Sei-i Taishogun (barbarian-subduing General) from the Emperor (a designation that actually signified the consignment of ruling authority by the Emperor) and consolidated all administrative authority in his own hands. He established the Bakufu (literally "tent-government", it signified rule by a military family) in Kamakura in the Kanto Region, so the following period is called the Kamakura Period (1185–1333).

After the death of Yoritomo, power within the Kamakura Bakufu shifted to the Hojo Family. Japan was attacked twice by Mongol forces, in 1274 and in 1318, but they were driven off by the Hojo.

From 1336 to 1392, the court was divided in two, the Southern Court and the Northern Court, accompanied by continuous struggle for the Imperial throne (the age of the Northern and Southern Courts). Emperor Godaigo of the Southern Court made an unsuccessful attempt to recover political authority from the Bakufu, but General Ashikaga Takauji, who was sent to oppose the Emperor, grasped power for himself. Takauji was awarded the Sei-i Taishogun title by the Northern Court and established his own Bakufu in Kyoto, ushering in a period called the Muromachi Period (1336–1573). The Muromachi Period is considered to be an age of great progress by Japanese culture, but it was also a period of political instability.

The Bakufu could not control the powerful regional clans (groups with military control of specific regions) and as their power gradually expanded, in the Sengoku Period (1467–1568), an age of nationwide struggle for power began.

For the powerful regional clans, flood control and agricultural land development were major challenges which had to be met to expand their power, so they conducted land development throughout Japan. To take Takeda Shingen, who was the leader of a powerful regional clan in present-day Yamanashi Prefecture as an example, he systematically developed agricultural land at the same time as he worked to improve rapidly flowing rivers by constructing groins, open levees and planted buffer zones, etc. to prevent disasters along rivers.

During this age of conflict, three men played outstanding roles in the process of unification: Oda Nobunaga, Toyotomi Hideyoshi, and Tokugawa Ieyasu and this period is called the Azuchi-Momoyama Period (1573–1600). Nobunaga began unification by destroying opposing forces and was later forced to commit suicide during a rebellion led by his retainers. Following Nobunaga, Hideyoshi grasped power, finally uniting Japan in 1590.

Construction technology that was developed in the Sengoku Period achieved rapid progress during this age. Nobunaga constructed Azuchi Castle, the largest such structure built in Japan at that time, together with its surrounding castle town. Hideyoshi mobilized his planning capacity and organizational skills to construct Osaka Castle, Fushimi Castle and others, and the Ogura-Ike Pond improvement project he carried out in the south of Kyoto is more properly described as an integrated development project than as a mere river improvement project.

After Hideyoshi died, the great task of national unification was taken over by Ieyasu.

1.2.5 The Edo Period (1600 to 1868)

Previously mandated governance of Edo (now Tokyo) by Hideyoshi, Ieyasu destroyed all forces opposing him, and in 1603, he established his Bakufu in Edo, thus beginning the Edo Period. The Bakufu established the Bakuhan System (an administrative structure led by the Bakufu), and despite famine and financial problems, it maintained peace for approximately 260 years.

Portuguese and Spanish missionaries had been visiting Japan since the Sengoku Period, but because the Tokugawa Bakufu prohibited Christianity and sharply restricted foreign trade, a policy of national seclusion was enforced, with contact with western European countries limited to relations with the Dutch in Nagasaki, resulting in the Bakufu monopolizing trade.

Foreigners began to approach Japan frequently in the last half of the Edo Period. In 1853, a fleet of American naval vessels led by Commodore Perry visited Japan, severely shocking the nation. The following year, Japan was forced to sign an unequal treaty called the Treaty of Peace and Amity between the Empire of Japan and the United States. This exposed the incompetence of the Bakufu leaders, eventually sealing their fate.

During the last ten years of the Edo Period, which is called the Bakumatsu (end of the Edo Period), many confrontations with external forces were accompanied by repeated conflict between pro-Bakufu and anti-Bakufu factions within Japan. During this period of disorder, the fifteenth leader of the Bakufu, the Shogun Tokugawa Yoshinobu, returned political authority to the Imperial Court, ending the Edo Period. This is called the Taisei Hokan (return of Political Power to the Emperor) (1867).

The Edo Period was an age of peace and also an age of Japanese development. The Bakufu and the Hans[2] devoted their greatest efforts to flood control and agricultural land development projects. It is reported that about 70% of water use systems now used for paddy field agriculture, such as irrigation ponds and irrigation channels, were completed during the Edo Period. Castles constructed by powerful regional families were often located on plains and urban development projects to build cities were carried out around castles. A well-known development, typical of the Edo Period, was the construction of Edo Castle in the center of present-day Tokyo, accompanied by the diversion of the giant Tone River that flowed into Tokyo Bay to the Pacific Ocean in the east, to provide flood control and river transport. Another was a project to introduce a municipal water system to the 1 million residents of Edo City.

1.2.6   The Meiji Period (1868 to 1912)

Under the Emperor Meiji, the government removed the powers of the former Bakufu to begin reforms to create a modern country, changed the name of Edo to Tokyo, and in 1869, moved the capital from Kyoto to Tokyo. In 1871, it abolished the hans and established prefectures.

To modernize Japan, the government raised the nation's capabilities by employing scholars, teachers and engineers from the industrialized countries of Europe and America and by sending young talented people to the industrialized countries to study. In this way, the systems of a modern state were created in a short time and in 1881, the Constitution of the Empire of Japan was promulgated, based on modern constitutional principles.

In the latter half of the Meiji Period, Japan expanded its power to the continent through the Sino – Japanese War of 1895 and the Russo-Japanese War of 1905.

Japan had developed its industries led by the government and modeled on advanced countries in Europe and America, achieving rapid industrialization in a short time. It began with spinning and weaving, then progressed to steel, chemicals,

---

2 Han: These were financially independent regional administrative organs led mainly by regionally powerful families and their retainers. There were well over 200 hans in Japan and the Bakufu interfered in their affairs only when a problem arose. The Bakufu ordered each han to carry out a variety of projects according to its scale.

shipbuilding, and armaments, thus completing its industrial revolution by the early 20th century.

Foreigners provided guidance with the provision of public infrastructure in many areas including flood control, railways, ports and harbors, sabo, agriculture, and so on, with the priority on railways and flood control structures in particular. In 1889, trunk railways began service between major cities, and in 1896 the River Law was introduced and the implementation of projects focused on preventing floods began throughout Japan. In Yokohama, Kobe, and other port cities, the creation of modern cities was accompanied by the construction of modern water supply and sewerage systems to fight infectious disease brought by foreign ships. The creation of this public infrastructure made a big contribution to the modernization of Japan.

1.2.7 The Taisho, Showa, and Heisei Periods (1912 to present day)

1.2.7.1 Before World War II (1912 to 1945)

During the Taisho Period (1912–1926), the first political party cabinet was established and democratic government advanced. This democratic period is called Taisho Democracy.

Japan participated in World War I as an ally of Great Britain. This war greatly expanded Japan's economy, but its end brought a slump. Furthermore, in 1921, the Kanto Earthquake (Mg. 7.9) devastated the capital, worsening economic problems and casting a shadow over Taisho Democracy.

It was followed by the Showa Period (1926–1988), which began with a financial crisis. Under these conditions, the road to military expansion in Asia was opened, resulting in the Manchurian Incident, followed by the Shanghai Incident that escalated into the Japan – China War.

At the same time, relations with the U.S.A. deteriorated. In 1941, Japan became involved in World War II when it started a war with the U.S.A. Although Japan scored early victories on various fronts, it was forced to gradually retreat, until the atomic bomb attacks on Hiroshima and Nagasaki in 1945 forced it to surrender by accepting the Potsdam Declaration.

It was unavoidably necessary to either slow or halt many public infrastructure projects during this period because of a shortage of funds or labor. But it is noteworthy that the Kanmon Undersea Tunnel (3.6 km, 1944) linking Kyushu and Honshu was completed during this period.

1.2.7.2 After World War II (from 1945)

On August 15, 1945, Japan was defeated and occupied by the Allies. In the following year, the Constitution of Japan was promulgated, followed by the introduction of democratic policies. Later, in 1951, the San Francisco Peace Treaty was signed, restoring Japan's autonomy.

Postwar Japan began its voyage in the midst of post-war chaos and repeated earthquake and typhoon disasters. The government implemented various policies, positioning disaster prevention measures and the provision of food and energy as challenges to be overcome in the early restoration process. Through the united efforts of government and the private sector, and under the effects of the Korean War (1950 to 1953), the Japanese economy recovered at an astonishing rate, with its GNP growth rate during the nineteen-sixties exceeding 10%. In 1970, Japan's GNP was 2.45 times its 1960 level and it was the world's second largest economy. (The so-called "rapid economic growth" period.)

The population in Japan had exceeded 100 million people. The rapid economic growth was accompanied by a concentration of the population in cities, resulting in a variety of urban problems, including water and air pollution, soaring land and housing prices, uncontrolled development and a shortage of water. In the1970s, a series of oil shocks occurred, so from about 1975, Japan's economy entered a period of stable growth.

In 1985, America's floating exchange rate system was adopted, and the Japanese economy entered the so-called "Bubble Economy" stage, marked by an accelerated rise in asset prices, which outstripped actual economic growth. The Bubble Economy collapsed at about the same time as the start of the Heisei Period (1989), resulting in severe economic confusion as a series of banks and large corporations went bankrupt. This was followed by a long deflationary period, the first to occur in postwar Japan and one that continues to the present day.

Civil society matured under postwar democracy, but, on the other hand, environmental problems appeared to expand into global problems such as global warming and as Japan enters the age of low birth rates and an aging society, the search for reforms needed to achieve a sustainable future continues.

During this period, the provision of public infrastructure encompassed roads, rivers, dams, railways and many other fields, and was carried out with investment of funds on a scale unprecedented in Japanese history. During the post-war rehabilitation period in particular, dams made great contributions to preventing disasters and ensuring energy and food supplies. Then, after the start of the rapid economic growth period, they helped to prevent disasters and to supply energy and water for municipal and industrial use, supporting rapid growth and earning this period the name, "Age of Dams."

Provision of public infrastructure after the start of the age of stable growth has been required to prepare for the future, but worsening financial conditions and environmental and social problems sharply reduced its pace.

1.3 OUTLINE OF THE HISTORY OF DAMS IN JAPAN [3, 4]

In the Yellow River area, Nile River area, Indus River area, Tigris-Euphrates area and other cradles of civilization, many traces of ancient water-use structures have been discovered, revealing a close link between civilization and water. There are many hypotheses concerning the world's oldest dam, but it is assumed to be a dam dating back to 2950 B.C. and whose remains are located in Memphis on the banks of the Nile River.

During the Spring, Autumn and Warring States periods in China, each country worked to expand its cultivated land. A remaining record reports that the Governor of Chu constructed a reservoir called the Anfengtang on land about 50 km south west of the present city of Huainan in Anhui Province in 6th century B.C. This reservoir is still in use today.

Irrigation ponds used to irrigate paddy field rice production in Japan (See p. 21, [1]) were introduced from China through the Korean Peninsula, and have reportedly been constructed since about the 3rd century. It is described as a major project in the Nihon-shoki (Chronicle of Japan) and the Kojiki (Record of Ancient Matters). The Sayama-ike Pond that is now in use (Osaka Prefecture, early 7th century) has been reconstructed several times since it was first built. The embankment of this irrigation pond is made of compacted alternating layers of soil and leaves.

When ancient Rome conquered various parts of Europe, it constructed many dams[3] along with aqueduct bridges, etc. These include dams such as the Proserpina Dam (Figure 4.4.1) constructed in the 2nd century in what is now a suburb of Merida in western Spain, and which remains in use after a series of reconstructions over a period of 17th to 18th centuries. Cement has been used to construct dams since Portland cement was developed in England in 1824. By permitting the construction of concrete gravity dams and arch dams, it caused a major revolution in dam technology, mainly in England and France. In the late 19th century, dams were constructed in various parts of England to supply water required by the advance of urbanization.

In Japan, modern western technology was introduced for dam construction in the Meiji Period. The Port City of Kobe that was suffering an epidemic of cholera did not have rivers from which enough water could be drawn, so the Gohonmatsu Dam (PG, 33.3 m, #94), otherwise called Nunobiki Dam, was constructed as Japan's first concrete dam to obtain water for the municipal water supply system. Beginning in the Meiji Period, modern civil engineering technology was applied to construct water supply dams and hydroelectric dams, in addition to agricultural use dams, bringing major changes to both industry and the daily lives of the people.

As part of the New Deal introduced in the U.S.A. in 1933 to deal with the Great Depression of 1929, dams were constructed in various parts of the country. The TVA (Tennessee Valley Authority) was established to construct multi-purpose dams at 32 locations as part of integrated development of the Tennessee River. These dams were economically very successful because they contributed to industrial and agricultural development by generating electricity, supporting irrigation, and controlling flood discharge. In the west, the US Bureau of Reclamation developed the Colorado River in southern California, which is an arid region, and in 1935, completed the 221 m high Hoover Dam that played a revolutionary role in water usage. This dam can be described as a product of combining all dam technologies and has served as the model for large dam construction around the world.

At the end of World War II in 1945, controlling floods, increasing electric power production, and supplying water to boost food production in Japan were challenges overcome by carrying out post-war restoration. In Japan, this challenge was tackled by comprehensive national land development, centered on dams: an approach modeled

---

3 Dams constructed in Ancient Rome: were made by supporting two parallel masonry walls with a hard core (the impervious part) between them, with soil or stones from both the upstream and downstream sides.

on the American TVA. The construction of multi-purpose dams, the purposes of which include flood control that were partially constructed following pre-war surveys but suspended when the war started, became the core projects in comprehensive national land development in the post-war years. The Sakuma Dam (PG, 155.5 m, #82) on the Tenryu River that was constructed as part of the promotion of electric power development, incorporated a construction method that introduced large American-made construction machinery, creating a major technological revolution in the construction industry and establishing present-day dam construction technology. It is not difficult to imagine that in Japan at that time, dams were welcomed amidst great expectations that they would help create a democratic society (p. 58, See box).

From the rapid growth period that started in the late1950s to the stable growth period, in response to public demand, multi-purpose dams and hydroelectric dams (including pumped-storage hydropower generation dams) continued to be constructed, mainly to control floods, supply municipal and industrial water and generate electricity, and their scale was increased. In this way, Japan became ranked among the top countries possessing large dams (Table 1.3.1), but the reservoir capacity that represents storage capacity totals 22.2 billion m³, which does not equal the storage capacity of a single Hoover Dam (40 billion m³) [6].

However, in recent years, people in Japan have begun taking a critical attitude to dam projects, because of deteriorating financial conditions, slowing increase of water demand, as well as the environmental and social impacts of dams.

This includes public doubts concerning the way dam projects are implemented: opposition to the construction of the Nagaragawa Estuary Barrage (Weir, #86) during the 1990s for example [7]. From an international perspective, almost at the same time (1994), Director Beard of the Bureau of Reclamation of the U.S. Department of the Interior stated, "the age of dam construction has ended in America [8]", a statement that caused world-wide ripple effects.

In this way, during the 1990s, pro- and anti-dam construction factions held opposing views. To resolve this dispute, in 1998, the WCD (World Commission on Dams)

Table 1.3.1 Top 10 countries by number of large dams.

	Country	ICOLD World Register of Dams 1998	Other sources	Percent of total dam
1	China	1,855	22,000	46.2
2	United States	6,375	6,575	13.8
3	India	4,011	4,291	9.0
4	Japan	1,077	2,675	5.6
5	Spain	1,187	1,196	2.5
6	Canada	793	793	1.7
7	South Korea	765	765	1.6
8	Turkey	625	625	1.3
9	Brazil	594	594	1.2
10	France	569	569	1.2
11	Other Countries	7,572	7,572	15.9
	Total	25,423	47,655	100

Source: WCD [5].

was formed with representatives from both sides, and in 2000, the WCD Report, Dams and Development: a new framework for decision-making was published [5, 9].

In this chapter, section 1.5 touches on points made in the WCD Report and on the state of the construction and management of dams in Japan. Firstly, the characteristics of the national land related to dams are described in section 1.4. The section explains why Japan, which is a long and narrow country, is one of the world's leading large dam builders and why it uses many dams.

1.4   CHARACTERISTICS OF NATIONAL LAND AND DAMS IN JAPAN

1.4.1   Characteristics of the national land and water usage by dams

The Japanese Archipelago lies at the eastern extremity of the Asian continent, where it extends from northern latitude 45°30′ to northern latitude 24°. Its climate varies from temperate in the north to sub-tropical in the south. Overall, it is part of the Asian monsoon zone and is an island chain covered by steep mountains in a region of dynamic mountain-building activity. As a consequence, conditions in Japan make it susceptible to disasters such as floods, droughts, torrential rainfall, landslides, debris flows, typhoons, earthquakes, tsunamis, volcanic activity, snow and high tides.

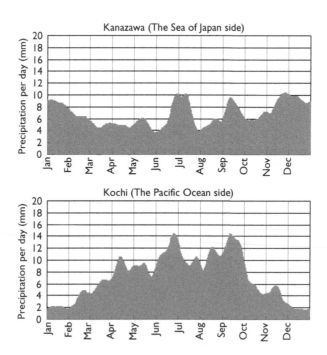

Figure 1.4.1 Typical precipitation pattern in a year.

Source: Rika Nenpyo [10]. Based on the value of the average year by the data of 30 years (1971–2000).

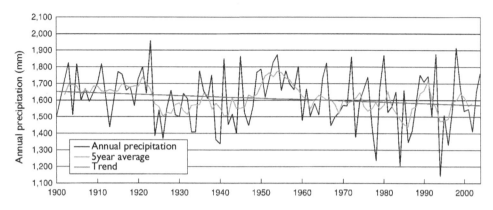

Figure 1.4.2 Trend of annual precipitation in Japan.

Source: MLIT [11].

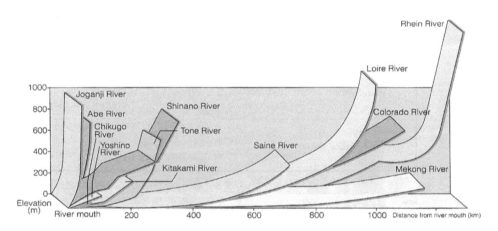

Figure 1.4.3 Comparison of river slopes in Japan and other countries.

Source: MLIT.

Characteristics of the climate are heavy rainfall and its irregularity. The average annual precipitation in Japan is 1,718 mm, which is high for a temperate zone. In every part of Japan, precipitation varies widely from year to year. While there are periods of light rainfall, there are others when massive quantities of rain fall in a short period: when typhoons approach and towards the end of the rainy season (Figure 1.4.1). Heavy snowfalls occur during the winter, mainly on the Japan Sea side of the country.

Abundant rain falling from spring to early autumn and the oceanic high temperature and humidity in the summer support Japan's rice paddy agriculture, which permits the country to support a high population. Hydropower production has thrived

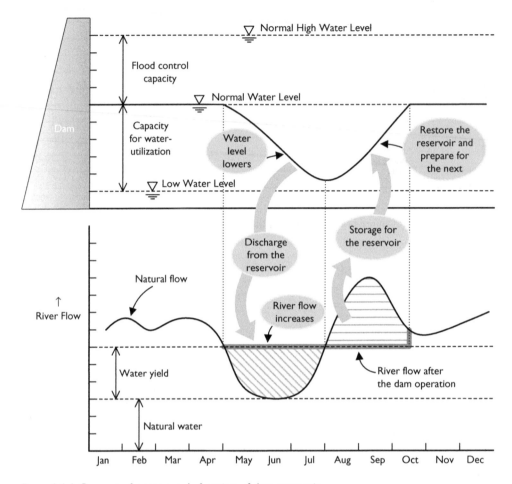

Figure 1.4.4 Concept of water supply function of dam reservoir.

thanks to heavy annual precipitation that includes snowfall. Conversely, Japan suffers from frequent flood damage, forcing it to prioritize flood control.

The average annual precipitation in Japan is almost twice the global average, but fluctuation in annual precipitation from year to year results in a large gap between years of heavy precipitation and years of light precipitation (Figure 1.4.2).

In recent years, rainfall in Japan has tended to be light, but annual rainfall has fluctuated greatly with concentrated torrential rainfall tending to occur frequently. Typhoons have occurred more frequently in August and September during the past 20 years, resulting in heavy rainfall [11].

Turning to the topography, mountains cover 70% of Japan, with 3,000 m ranges running up the center of Japan. Lowlands with an elevation of 18 m or less above sea level cover only 18% of the national land. Consequently, the rivers originate at high elevations and are not far from river mouths, resulting in the gradient of Japan's rivers

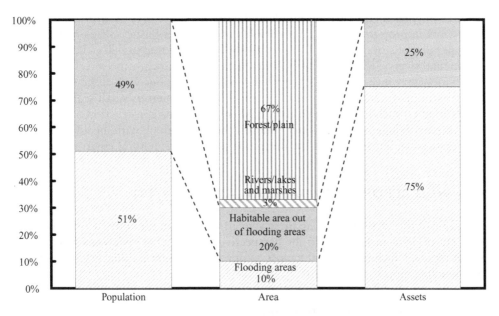

Figure 1.4.5 Utilization of national land – concentration of population and assets on alluvial plains.
Source: MLIT [12].

being far steeper than that of rivers in other countries. The upstream river gradient is much steeper than the downstream river gradient (Figure 1.4.3).

Dam reservoirs provide extremely effective flood discharge control functions enabling them to play a vital role in utilizing the rivers with such characteristics to ensure stable supplies of water for agricultural users, municipal water supply systems and industrial water users, while allowing flood runoffs to be dealt with safely (Figure 1.4.4).

1.4.2   Characteristics of regions surrounding dam reservoirs and rivers

Japan imported rice paddy agriculture in ancient times. Paddy fields that form hamlets with high population density and provide high productivity based on a labor-intensive form of agriculture, spread from alluvial plains downstream to plains in the valleys upstream. The valleys provided important transportation routes that supported the economy and culture of Japan. In modern times, 50% of its population and 75% of its assets are concentrated on the alluvial plains that cover only 10% of the national land [12]. River water is used to maintain the country's high productive capacity, which supports the large population and the national land is used under the influence of heavy inundations[4] (Figure 1.4.5).

---

4 Inundation: means a flood discharge overflowing a river bank or levee. When a flood discharge runoff begins on a certain river, the region reached by the inundation is called the inundation area.

As a result of the natural environment described in section 1.4.1, its topographical features in particular, irrigation ponds and dams create reservoirs with a storage capacity low in proportion to the dam heights in each river basin, so that construction costs per unit of storage capacity are relatively high compared to that of large dams on continental rivers. However, constructing dams that require advanced technology and heavy monetary investment has provided beneficiary regions with sufficient benefits. River basin management performed by constructing numerous small scale reservoirs of this kind on each tributary in a river basin results in the construction of many dam reservoirs with high reservoir water rotation rates[5] and along with locating them in a temperate zone, has minimized the impact of dams on the natural environment.

1.5   WCD REPORT AND DAMS IN JAPAN [5, 9, 12]

The topic of this section is the Japanese view of the state of the construction and management of dams in Japan from the perspective of the global review provided by the WCD report, which is seen as the most comprehensive international review of trends surrounding dams in recent years.

1.5.1   Outline of the WCD report

To resolve the worldwide conflict on the issue of dams that began in the 1980s, the WCD (World Commission on Dams) was formed at the initiative of the World Bank and the IUCN (International Union for Conservation of Nature and Natural Resources), an environmental NGO. The 36 countries and 68 organizations that are members have held international discussions on the development of water resources by dam projects. The results of their deliberations are, as stated in section 1.3, compiled as WCD reports.

We believe there can no longer be any justifiable doubt about the following:

- Dams have made an important and significant contribution to human development, and the benefits derived from them have been considerable.
- In too many cases an unacceptable and often unnecessary price has been paid to secure those benefits, especially in social and environmental terms, by people displaced, by communities downstream, by taxpayers and by the natural environment.
- Lack of equity in the distribution of benefits has called into question the value of many dams in meeting water and energy development needs when compared with the alternatives.
- By bringing to the table all those whose rights are involved and who bear the risks associated with different options for water and energy resources development, the conditions for a positive resolution of competing interests and conflicts are created.

---

5 Rotation rate: is defined by the following equation and represents the number of times the water in a reservoir is replaced annually as it is stored and discharged from the reservoir. Rotation rate = (total water flowing into a reservoir per year)/(effective storage capacity of the reservoir).

- Negotiating outcomes will greatly improve the development effectiveness of water and energy projects by eliminating unfavourable projects at an early stage, and by offering as a choice only those options that key stakeholders agree represent the best ones to meet the needs in question.

This achievement by the WCD offers many suggestions on ways to carry out sustainable development, particularly in developing countries that are now also constructing large dams.

1.5.2 Japan's dams from the perspective of the global review

Not all the contents of the WCD Report necessarily conform to circumstances of dams in Japan, but many points in the global review concerning project effectiveness and their environmental and social impacts should be considered as indications of the situation in Japan.

Dams are constructed and managed in Japan by carefully studying the natural and other conditions of the region surrounding individual dams, which are smaller than the giant dams constructed in continental nations. Consequently, we are considering ways to avoid the severe environmental and social impacts of large continental dams. This is because:

1. the concentration of highly productive irrigated rice agriculture and the population on flood plains has required Japan to operate many small dams in mountainous regions outside these flood plains,
2. and measures for resettlement of the inhabitants of reservoir areas (see the main text on p. 103–) and water use regulation in downstream areas (see the main text on p. 113–) have both functioned extremely well.
3. A series of systems has been established, including environmental impact assessments, measures to avoid or reduce environmental impact, post-construction environmental monitoring, post-project evaluation of managed dams and the reflection of the views of concerned residents, in the enactment of River Improvement Plans (see p. 143, [11]).

These measures can become a model for dam construction projects and dam management around the world. Chapter 3 describes the state of these innovations.

The impacts of dams on natural environments are diverse and complex. The WCD Report points out that there is not enough knowledge to be able to assess the impact of dams appropriately. Therefore, even in Japan, dams have been selected when there is no feasible alternative proposal considering the need for its construction, its purpose, its effectiveness and so on.

Items concerning the impacts of dams on society that are pointed out in the WCD Report are all addressed properly in Japan where economic assessments have been possible. However, it is necessary to think about highlighted points such as the psychological stress on those living in the area that will be submerged—stress caused by delayed projects or a long time from the preliminary study to the start of construction—and also the impact on the local culture: the loss of rural communities for example.

The WCD Report attempts to evaluate the direct benefits of dam projects plus their financial and economic aspects, presenting information from many perspectives that should be of concern to Japan.

1.6   OUTLINE OF THIS WORK

This work deals with the challenges of clarifying the roles and impacts of dams, based on the background described above under three topics: roles that dams have played; negative impacts of dams and responses to these; and future roles of dams.

1.6.1   Roles that dams have played

This report begins in Chapter 2, which clarifies the roles that dams have played in the past, divided broadly into five roles.

a   Throughout history, and up to the Meiji Period in particular, agricultural dams played a role in the development of rice agriculture, which has maintained the economy and the population of Japan. From the Kofun Period through the Edo Period and up to the end of World War II, irrigation ponds and agricultural dams constructed as part of new rice field development continued to play an important role, and since the war in particular, they have provided water to support the advance and transformation of agricultural production.

b   Shortly after the opening of Japan in the Meiji Period, dams acquired a new role as suppliers of water to municipal water supply systems to improve sanitary conditions in cities: mainly port and harbor cities.

c   Through the Meiji, Taisho and early Showa Periods (1868–1945), hydropower dams produced more than half of the electric power for daily life and industry in Japan during that period. This period, when most electric power was hydro power followed by thermal power (see p. 44, [15]), continued for about half a century, from 1911 to 1962.

d   Multi-purpose dams with flood control capabilities played a role in restoring the country following World War II. Comprehensive National Land Development Projects centered on multi-purpose dams not only ensured safety by controlling floods, but also responded to shortages of food and energy by supporting increased food production and developing electric power sources along with flood control.

e   Dams provided municipal and industrial water in response to demand created by the concentration of industry and population in the large metropolitan regions during the rapid economic growth period (after 1960s) and controlled floods to strengthen the flood-safety of cities. As the share of electric power supplied by nuclear power plants increased during this period, pumped-storage hydropower dams came into use as a way of meeting peak demand for electric power.

In this way, river development to meet the needs of each period have formed the present water cycles on national land, against the backdrop of historical conditions. This means that modern Japan has benefited from a history of water and land use by previous generations in Japan, mainly by constructing dams. The above is the bright side of the dam story.

1.6.2 Environmental and social impacts of dams and response to these impacts

In contrast to the roles played by dams that are described in Chapter 2, dams have had considerable impacts on social conditions in dam reservoir areas.

Chapter 3 begins with discussions of the impacts of dams and measures taken in response to these impacts from the perspective of public opposition to dam projects. Then it describes each of the following problems and the measures taken to resolve them.

a Impacts of dams on social conditions in reservoir areas and responses to these impacts: measures for reservoir area development focused on the resettlement of the inhabitants.
b Impacts of dams on flow regimes on downstream rivers and response to these impacts: regulating the use of river water and ensuring the river maintenance flow (see p. 118, [4]).
c Sedimentation in reservoirs, cold water discharge, prolonging of turbidity, eutrophication phenomenon and other impacts on biological habitats and response to these impacts.
d Response to the growing citizens' movements in opposition to dam projects.

Parties that construct and manage dams take the following measures where possible, not only to prevent possible environmental and social impacts, but to minimize these impacts by avoiding or mitigating them and providing compensation when they do occur.

i   Improving systems, legislative measures, apportioning the roles of all stakeholders and cooperating with them.
ii Harmonizing the interest of all stakeholders.
iii Performing research and developing technologies.

In the face of growing citizens' movements, they work hard to fulfill their responsibi-lity to explain the need for projects and to increase the transparency of assessments. Under the River Law as revised in 1997, legal procedures that reflect the views of concerned residents in the enactment of River Improvement Plans have been established. Under River Improvement Plans prepared as trial proposals by the Ministry of Land, Infrastructure, Transport and Tourism (MLIT), the impacts on the environment of multiple alternate proposals prepared at the planning stage, based on the desired form of the river, are comparatively assessed. In this way, the responses to the dark side of dams constructed in recent years have been clarified.

1.6.3 Future roles of dams

Chapter 4 tackles the future roles of dams based on future prospects including the international perspective as three roles:

a Roles in ensuring agricultural use water to guarantee food supplies in response to the rising world population.

b   Roles in ensuring hydropower to increase national production and to recycle energy.
c   Roles in ensuring and expanding flood control and water supply functions to deal with the growing range of fluctuation of floods and droughts under the effects of global warming.

This work concludes that it is, therefore, necessary to promote proper management and maintenance, and effective use of existing dams, and that engineers should improve their professional capabilities to support dams that are constantly becoming more sophisticated and contribute to international cooperation.

Chapter 2

Roles played by dams as seen in the history of water use in Japan

Human water use undoubtedly began with drinking water. Later, as our ancestors began farming, irrigated farming in particular, the quantity of water they used increased greatly and they began constructing dams.

Dams are now broadly classified as water use dams and flood control dams according to their purpose, and water use dams are further categorized as irrigation dams, municipal/industrial water dams (municipal water supply dams and industrial water supply dams) and hydropower dams. A single dam that is used for two or more of the above purposes is called a multi-purpose dam.

This chapter begins by examining the role of irrigation dams, which were probably the first type of dams used, with a long history of water usage.

2.1 THE DEVELOPMENT OF IRRIGATION AND IRRIGATION DAMS

As shown by Figure 2.1.1, since the ancient Kofun Period, irrigation ponds[1] have played an important role in the spread of rice farming, and by the Edo Period, agriculture in Japan supported a population of 30 million people. Irrigation dams incorporating modern engineering have played a major role in guaranteeing stable irrigation, beginning during the period when food production rose sharply after the end of World War II to the present.

2.1.1 The history of methods of obtaining irrigation water and the roles of irrigation ponds [14–17]

2.1.1.1 Development of rice agriculture and history of irrigation ponds

Japan reportedly changed from a hunting-gathering society to an agricultural society about 300 BC. Rice cultivation, which was brought to northern Kyushu island by migrants from the Eurasian continent, rapidly spread eastwards, reaching the northern tip of the main island by about the 3rd century.

---

1 Irrigation pond: "TAMEIKE" in Japanese. Reservoirs formed by earth-fill dams that were constructed for irrigation purposes. Almost all irrigation ponds are small saucer-shaped ponds with a storage capacity between a few thousand and 10,000 m³ of water.

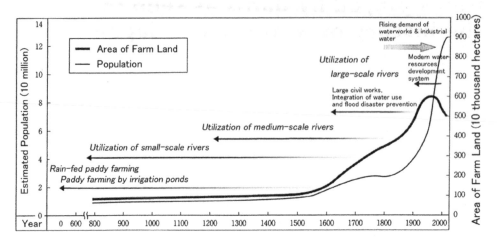

Figure 2.1.1 Changes in farm land area and population of Japan [13].

Source: National Land Agency.

The early age of rice cultivation relied on rain falling on land where wetlands or springs, etc. could be utilized. After the end of the Yayoi Period, when iron tools were first used, irrigation, that is artificial water supply, was started when it became possible to construct small channels to take in and transport water from small rivers.

As primary irrigation, water was probably taken from small rivers naturally. This was replaced by water supply stabilized by weirs. On new paddy fields and other places where the quantity of water necessary to grow rice could not be ensured, supplementary irrigation with water stored in irrigation ponds appeared for the first time. These technologies were probably introduced by migrants.

It is recorded in the Nihonshoki (completed in 720)[2] etc. that the Yamato Court[3] constructed irrigation ponds by applying the nation's total energy. It is estimated that the Sayama-ike Pond, which is still in use, was constructed in Osaka Prefecture in the early 7th century based on a study of the age of pottery discovered there. Also noteworthy is that the height of the dam body has been raised several times [18].

The reforms that established the centralization of administrative power by the Emperor and aristocrats (645) included measures to nationalize people, land and water. The expansion of agricultural land and water usage systems were aggressively promoted by the central government. The Manno-ike Pond in Kagawa Prefecture was built during this period (about 700) [19].

Paddy field management by the state collapsed as private ownership of land by the aristocrats matured. Later in the middle of the 8th century, it was replaced by the administration of paddy field within each territory by landowners. Then, as the age of rule by the warrior class arrived in the 12th and 13th centuries, water resource

---

2 Nihonshoki: refer to 1.2.2.
3 Yamato Court: refer to 1.2.2.

Table 2.1.1 Major irrigation ponds constructed by the 15th century.

Name of irrigation ponds	Pref.	Height (m)	Year of completion
Sayama-ike	Osaka	8	The early 7th century
Manno-ike	Kagawa	10	About 704
Nagayuki-ike	Kagawa	10	The early 11th century
Daimon-ike	Nara	18	1128
Ootani-ike	Kagawa	13	1470

Source: K. Yukawa [20].

development transcended the framework of private management systems. From then until the Sengoku period, in Western Japan, small irrigation ponds were constructed in districts suffering from water shortages, agricultural technology, such as introducing double cropping in paddy fields and breeding of crops developed steadily. Table 2.1.1 shows the major irrigation ponds constructed until this period.

2.1.1.2   The establishment of river irrigation and irrigation ponds in the Sengoku and Edo Periods

Throughout the Sengoku and Edo Periods, under the protection of the central government and regional feudal lords, irrigation systems appeared across Japan as intake weirs were constructed in the middle reaches of large rivers and long large irrigation canals were constructed to transport water, thereby stabilizing the supply of irrigation water. New paddy field development driven by progress in flood control and irrigation technologies expanded to the previously-unexploited delta zones, which are the major flood plains of rivers, and these new agricultural areas were incorporated into the networks of irrigation systems supplied by large rivers.

In the Edo Period, the widespread development of new paddy fields supplied by irrigation ponds began, mainly in the Hans[4] of western Japan. This was supported by the spread of irrigation pond construction technologies, which were based on castle-building and mining technologies.

These new field development projects expanded the total area of cultivated land from an estimated 2.06 million hectares in the beginning of the 17th century (including 1.5 million hectares of paddy field land) to 4.13 million hectares (including 2.27 million hectares of paddy field land) by the end of the 19th century: an increase of about 2 million hectares (800,000 hectares of paddy field) in a little less than 400 years [21]. Prior to the Edo Period, irrigation ponds with a supply area of 2 hectares or more, and for which the construction date is specified, existed in about 17,000 locations, and these irrigation ponds had a total reservoir capacity of about 700 million m³ and a supply area of about 220,000 hectares. The major irrigation ponds constructed during this period are shown in Table 2.1.2.

---

4 Han: refer to p. 7 [2].

Table 2.1.2 Major irrigation ponds constructed from the 16th to the 18th century.

Name of irrigation ponds	Pref.	Height (m)	Year of completion
Iwanabe-ike	Kagawa	13	The early 16th century
Shimo-ike	Ehime	18	The latter half of the 16th century
Iwase-ike	Kagawa	12	1592
Deguchi-ike	Yamaguchi	16	1601
Iruka-ike	Aichi	23	1633
Sakura-ike	Wakayama	24	1760s

Source: K. Yukawa [20].

The technological manual authored by Ohata Saizo[5] in the late 17th century, includes a method of calculating the capacity of irrigation ponds during new field development [22]. This suggests that at that time, reservoir capacities were calculated based on irrigation requirements, relying on experience and considering geographical conditions where new fields were developed.

2.1.1.3 Irrigation development and irrigation ponds from the Meiji Period up to World War II

During the Meiji Period when the feudal age was replaced by a modern unified nation, the existing irrigation management system under the Hans collapsed, thus requiring a new form of management.

Agricultural policies of the Meiji Government included the introduction of new irrigation technologies from the West, along with the importation of breeding animals and seed, etc. For example, the Dutch engineer Van Doorn, who was hired by the Meiji Government, introduced Western irrigation planning technologies in Japan to establish a national project plan (1879) to construct canals that would permit the opening of new land on the Asaka Moor in Fukushima Prefecture. This project, which was completed in 1882, was implemented as a land development project to provide agricultural land to members of the samurai class, who had lost their rights of occupation in the course of the collapse of the feudal system. A similar national government project was completed in 1885, at Nasunogahara in Tochigi Prefecture.

On previously cultivated land, arable land readjustment (land consolidation) projects to ensure irrigation water were also carried out in the Meiji Period, mainly among property owners, but the Land Consolidation Law was revised in 1905, and irrigation and drainage projects became main stream. Then, in 1923, central government rules were revised and implementation of large-scale irrigation projects with government

5 Ohata Saizo: 1642–1742. A Kishu Han (present Wakayama Prefecture) official who worked to develop water use.

Table 2.1.3 Large irrigation ponds from the late 19th century to the early 20th century.

Name of irrigation ponds	Pref.	Height (m)	storage capacity (10^3 m^3)	Year of completion
Mawari-ike	Kyoto	22.0	697	1880
Daijou-ike	Hyogo	29.0	944	1928
Taisho-ike	Wakayama	26.0	500	1915
Ookamidani-tameike	Tottori	28.0	1,096	1925
Shiote-ike	Okayama	28.5	1,250	1928
Jinnai-kamiike	Kagawa	29.7	750	1914
Honen-ike	Kagawa	32.3	1,590	1930
Ootani-ike	Ehime	27.0	618	1920

Source: Japan Dam Foundation [23].

Note: Extract dams whose height is over 20 m with a storage capacity of over 500,000 m^3 from the dams completed 1868–1930.

funding began. In this way, the central government replaced the landlords as the leading player in irrigation development.

Irrigation ponds and irrigation dams were constructed, upgraded and linked on and around medium and small rivers throughout Japan during this period, mainly in Western Japan (Table 2.1.3). Among these, the Honen-ike Dam (CB, 30.4 m, #110) that was constructed in 1930 was the first multiple-arch dam in Japan.

2.1.2 Reformation of cultivation and the roles of irrigation dams

2.1.2.1 Reformation of cultivation and the construction of irrigation dams after World War II

Rice cultivation in Japan was formerly done according to traditional cultivation, which began with transplanting in June and ended with the harvest in October and November. After the end of World War II, early season paddy rice cultivation technologies were developed and disseminated in order to avoid storm and flood damage caused by typhoons in September. These technologies permitted early completion of the harvest, adequate hours of sunlight and avoided cold weather damage in the Tohoku Region.

Because the spread of early cultivation dispersed and prolonged the season for which irrigation was necessary, irrigation dams played major roles in providing the newly-required water. During this period, irrigation dam construction was improved by adding methods based on modern soil engineering and concrete engineering, resulting in a rapid rise in the number of irrigation dams constructed with a height of 60 m or more.

Furthermore, paddy field improvement projects following World War II included the separation of water supply and drainage systems[6], improvement of field drainage by constructing underdrain systems and so on, and productivity was improved by using large agricultural machinery. Using these processes, the rising demand for irrigation water was met by constructing irrigation dams, etc. In recent years, the construction of irrigation dams, etc. has prevented drought damage on farmlands (upland farming), and permitted the development of diverse farming, including open-field vegetable, fruit and flowering plants, greenhouse cultivation, and the introduction of high value-added crops.

Facilities that take in river water have been improved and intakes have been unified in response to the agricultural reform. These were realized when water resources were improved by constructing multi-purpose and irrigation dams [17].

2.1.2.2   The present state of irrigation ponds and irrigation dams

The Ministry of Agriculture, Forestry and Fisheries (MAFF) conducted a fact-finding national survey of irrigation ponds and irrigation dams and prepared an irrigation pond ledger at the end of the 1996 fiscal year [24].

The survey revealed that, as shown in Figure 2.1.2, there are 210,769 irrigation ponds and irrigation dams throughout Japan. These include the numerous irrigation ponds in the Kansai Region which have been developed since ancient times and in the region surrounding the Setouchi Inland Sea where rainfall is low.

This survey targeted 63,591 locations with a beneficiary area of 2 ha or more and as a result, the state of irrigation ponds and irrigation dams has been classified as displayed in Table 2.1.4.

By identifying the year of construction, it can be shown that many dams and ponds were constructed before the Edo period and after World War II. About 15,000 irrigation ponds and irrigation dams, which account for about 23% of all such structures, have been constructed after World War II, but their total beneficiary areas and total effective reservoir capacities are equal to about half of the total: 49% and 47%, respectively. This shows that in recent years, the scale of irrigation ponds and irrigation dams has escalated dramatically.

These irrigation ponds and irrigation dams supply irrigation water to 1.226 million hectares of land. This is equivalent to 47% of all paddy fields, with an area of 2.6 million hectares, and to about 26% of all cultivated land, with an area of 4.7 million hectares.

In Japan, Land Improvement Districts[7] now promote regional water use. A multilayered management system has been established, with irrigation dams,

---

6 Separation of water supply and drainage systems: Replacing channels used to both supply and drain water with separate supply and drainage channels. Separating supply and drainage main canals prevents contamination of irrigation water with regional drainage water and eliminates the effects of rainfall runoff on irrigation water. The terminal separation of supplied water and drained water means building supply channels and drainage canals on paddy fields, thus permitting water management independent of other paddy fields. If supply and drainage are separated, drainage improves and repeat-use is prevented, thereby increasing water demand.

7 Land Improvement District: An independent group of farmers, established by law, to implement irrigation and drainage projects, etc. and provide an agricultural production infrastructure. There are now about 7,000 Land Improvement Districts in Japan.

Figure 2.1.2 Geographical distribution of irrigation ponds.

Source: MAFF [24].

irrigation ponds, canals and other major facilities managed by a Land Improvement District, small canals by hamlet level management organizations, and the water supplied directly to paddy fields managed by farm households.

Irrigation ponds and irrigation dams not only fulfill their original role of maintaining and improving agricultural productivity; they also store flood discharge in the capacity left by supplying water during irrigation periods, provide auxiliary functions that reduce flooding downstream, and create water environments that can provide urban residents with places for recreation and relaxation.

Efforts to use dams more effectively by raising dam bodies to increase reservoir capacity in response to rising demand for irrigation water, are taken accompanying the decline in irrigation areas resulting from changes in the state of agriculture after a dam has been completed, thus using the dam more effectively by switching its irrigation capacity to other uses, or using it for flood control, or as river maintenance flow, etc.

Table 2.1.4 Percentages of irrigation ponds and irrigation dams by the year of construction.

	The number of irriga-tion ponds and dams	Total supply area (hectare)	Active storage capacity (10^3 m^3)
The Edo Period former (–1868)	16,702 (27%)	220,862 (18%)	708,360 (24%)
Meiji Period (1868–1912)	10,211 (16%)	118,993 (10%)	279,460 (9%)
Taisho Period (1912–1926)	2,381 (4%)	43,927 (4%)	83,351 (3%)
Showa Period before World War II (1926–1945)	3,290 (5%)	89,777 (7%)	200,672 (7%)
After World War II (1945–present day)	14,868 (23%)	606,739 (49%)	1,441,394 (47%)
Unknown	16,139 (25%)	145,584 (12%)	286,475 (10%)
Total	63,591 (100%)	1,225,882 (100%)	2,999,712 (100%)

Note: Irrigation ponds and dams with a supply area of 2 ha or more.

Source: MAFF [24].

2.1.3   Examples of efforts to ensure a constant supply of irrigation water

2.1.3.1   Development of irrigation ponds in Kagawa Prefecture and the historical role of irrigation ponds [19, 25]

2.1.3.1.1   A group of irrigation ponds in Kagawa Prefecture

The annual precipitation in Kagawa Prefecture on Shikoku Island is approximately 1,150 mm, which is about two-thirds of the national average, a particularly low precipitation. The rivers are steep with short channels, so rainwater is immediately discharged into the sea, the normal flow of rivers are low and conditions for water use are poor. Originally, the productivity of paddy fields in this region was higher than that of paddy fields in low wetlands, as long as there were supplementary water resources. Therefore, since ancient times, irrigation water has been provided by constructing and raising the height of, irrigation ponds.

In the Edo period, priority was placed on rice cultivation as the foundation for the han's finances and farmers also sought the foundations of their livelihoods in rice cultivation. So new paddy fields were developed and irrigation ponds were also constructed to supply them with water. Records of rice production in a han located in the eastern part of the present-day Kagawa Prefecture during an approximately forty-year period in the first half of the 17th century show that production during that period increased by more than 30%, from 26,000 tons to 35,000 tons. It is thought that new field development accompanied by the construction of irrigation ponds contributed most to this growth.

Irrigation ponds were also constructed during and after the Meiji Period. After World War II, irrigation water was ensured by the construction of irrigation ponds, which developed 31.5 million m^3 of water resources: a quantity that corresponds to about 30% of all water resources developed before the war.

A cycle has occurred, whereby ensuring irrigation water by building irrigation ponds has spurred the development of new fields, causing new irrigation water shortages,

Presented by Kagawa Prefecture

Figure 2.1.3 Manno-ike Pond. (See colour plate section)

requiring the construction and improvement of irrigation ponds to overcome this shortage, and so on. As a result, more than 14,600 irrigation ponds, including the Manno-ike Pond, have been constructed or raised, thereby ensuring irrigation water and expanding rice cultivation.

However, a significant mitigation of water shortages in Kagawa Prefecture had to wait until 1974, when the central government irrigation project to divert and convey water from the Yoshino River with the Sameura Dam (PG, 106.0 m, #117) and other sources of water, began to supply water.

The degree of dependency of Kagawa Prefecture on irrigation ponds was as high as 70%, until this project began to supply water, and even now, it still exceeds 50% as irrigation ponds continue to function as major sources of water.

2.1.3.1.2   The Manno-ike Pond [26] (Figure 2.1.3)

The Manno-ike Pond is formed by an earth dam that blocks the Kanakura River in Kagawa Prefecture, has a history of more than 1,300 years since its construction, and is the largest irrigation pond in Japan. Constructed in the early 8th century and later destroyed by a flood in 818, the Manno-ike Pond was reportedly reconstructed by Kukai.[8] Later it was destroyed and reconstructed repeatedly and then abandoned after it collapsed in 1184.

More than 90 irrigation ponds were constructed or expanded during the Edo Period, and one of these was the Manno-ike Pond, which was reconstructed in 1631, about 450

---

8 Kukai: 774–834. One of the founders of Buddhism in Japan. Kukai was also famous for his calligraphy. Refer to 1.2.3

years after its previous collapse. Its restored reservoir capacity of 4.96 million m³ supplied a vast beneficiary area encompassing three counties and 44 villages, which produced 5,400 tons of rice: 15% of the total production of 35,000 tons in this region at the time.

Records show that even during the Edo Period, the dam body was raised to boost reservoir capacity several times, but since the Meiji Period, it has been raised once, in 1905, thereby boosting capacity to 6.6 million m³. Then, in 1930, water was conveyed from the Saita River and the dam was raised for a second time, increasing its capacity to 7.8 million m³. In response to a later drought, a plan to raise it a third time was proposed, and this was completed in 1959. Its present reservoir capacity is 15.4 million m³, freeing farmers from strict water allocation practices.

In recent years, the Manno-ike Pond has contributed to supplying water for daily life, in addition to agricultural use. During a severe drought in Takamatsu City in Kagawa Prefecture in 1973, 5,000 m³ of water per day was supplied by the Manno-ike Pond through a main canal of the Kagawa Irrigation System. Today, the surroundings of the Manno-ike Pond have been developed into recreational facilities, including a national park, thereby adding recreation and relaxation to its functions.

2.1.3.1.3   The Honen-ike Dam (Figure 2.1.4)

The Honen-ike Pond is an irrigation dam on the Kunita River in Kagawa Prefecture. This dam, which was constructed during the last years of the Taisho Period and first years of the Showa Period, the age in which concrete dam construction technology first appeared in Japan, was built using a revolutionary new method called the multiple arch dam. It was designed by Sano Tojiro (refer to 2.2.1.2.2), who also participated in the construction of the Gohonmatsu Dam in Kobe City, described below.

The beneficiary area of the Honen-ike Pond is an area that formerly suffered from regular droughts and where people struggled to obtain irrigation water, so they could cultivate only about 30% of the farmland with the existing irrigation ponds. Beginning in the middle of the Meiji Period, farmers switched to paddy farming that required more water, because they were under pressure from cheap imports of

Presented by MAFF

Figure 2.1.4 Honen-ike Dam. (See colour plate section)

sugar cane and cotton, crops that they had cultivated until that time. This further accelerated the shortage of irrigation water, so the region's farmers used groundwater obtained from wells. As a result, wells were constructed at 12,000 different locations by the end of the Taisho Period.

The Honen-ike Pond was constructed to overcome such shortages of irrigation water. Work started in 1926 and was completed in 1930, after four years of hard work by 150,000 people.

Leakage worsened as a result of deterioration of the dam body in the early 1980s, more than a half century after it was constructed, so between 1988 and 1993, it was repaired with great care to preserve its basic shape and to maintain the surrounding scenery in a pristine condition. In 1997, it became the first dam to be designated as a registered, tangible, cultural property.

The discharge of water from the Honen-ike Pond has become an event symbolizing the season and the environment surrounding the dam. In the downstream area, the waterside park has been improved, transforming it into an attractive place for hiking, which is visited by sightseers in pursuit of beautiful scenery, from Kagawa prefecture and other regions.

2.1.3.2   Development of irrigation by the Sannokai Dam in Iwate Prefecture [28, 29, 30]

2.1.3.2.1   An outline of water usage before the World War II

The beneficiary area of the Sannokai Dam (ER, 61.5 m, #13) is an area of approximately 4,000 ha of paddy fields in the basins of the Takina River and the Kuzumaru River, which are tributaries of the Kitakami River almost directly in the center of Iwate Prefecture.

Agricultural land development in this area began in the early 9th century, followed by the construction of irrigation systems on the Takina River called the Twenty-seven Weirs, laying the foundation for irrigation. It is estimated that about 822 ha of land was irrigated by the last half of the 17th century.

In the Twenty-seven Weirs district, chronic water shortages resulted in fierce water disputes along the upstream and downstream river and strict regulations on the width of coffering and levee materials used on the river. The further upstream, the greater the susceptibility to leaking of structures and materials that were selected to build the weirs. The weirs were constructed to allow water to flow into the downstream river, and even management standards for leak prevention work had been enacted. Because of frequent water shortages, water disputes occurred, on average, once every five years for about 300 years, ending in the early Showa Period, and serious disputes even claimed lives.

2.1.3.2.2   Construction of the old Sannokai Dam

To resolve this severe water shortage problem, in 1944, the Sannokai Irrigation Project with the (old) Sannokai Dam as its water source was designed to supply irrigation water to 2,852 ha of old paddy fields and to ensure a new supply of irrigation water to the 406 ha of new paddy fields.

In 1953, work on the dam and on a 21 km main canal was completed, thereby ending the water disputes. Upland fields, forests and wetlands that could not be

Presented by MAFF

Figure 2.1.5 Sannokai Dam. (See colour plate section)

developed until then because of the shortage of irrigation water were finally developed, contributing significantly to increasing yields of rice, vegetables and other crops.

2.1.3.2.3   Redevelopment of the Sannokai Dam (Figure 2.1.5)

Later a marked tendency for large quantities of water to be used for surface soil puddling,[9] etc. appeared as a result of land consolidation of paddy field land readjustment, further expansion of paddy field areas, and the introduction of cultivation machinery. Because the formation of well-drained paddy fields by new field development increased the quantity of irrigation water, returned-flow use by damming up drainage canals plus block rotation, which means allotting irrigation water successively to each district, were carried out.

Therefore, in order to resolve the existing shortage of irrigation water and ensure sufficient water to meet future needs, the Sannokai Irrigation Project, which included the redevelopment of the old Sannokai Dam (raising the dam body) and conveying water from the adjacent Kuzumaru River, began in 1979 and was completed in 2002. The completion of the New Sannokai Dam made a great contribution to the development of the region's agriculture by eliminating fear of droughts and ensuring the supply of irrigation water. It provides a stable supply of water to the large blocks of paddy fields with little labor, by using pipelines and automatic supply hydrants.

---

9 Soil puddling: spreading water in paddy fields and mixing in soil before transplanting to homogenize the paddy fields and prevent leaks.

2.2   THE GROWTH OF MODERN CITIES AND DAMS
FOR MUNICIPAL WATER

Modern cities with their large populations require public water supply systems to maintain hygienic conditions. Where there are neither wells nor springs to provide this water and it is unobtainable from rivers, the only choice is to construct dams to store water.

2.2.1   The start of modern municipal water supply systems in Japan

2.2.1.1   Waterworks during the Edo Period

From ancient times, humans formed small hamlets beside springs or small rivers from which they could easily obtain drinking water. Ancient remains confirm that during the Yayoi Period, wells were already being constructed in hamlets and water was used as a tool in daily life.

The first waterworks intended mainly to provide drinking water was the Kanda Waterworks constructed by Tokugawa Ieyasu in 1590. It was expanded along with the establishment of the urban functions of Edo (now Tokyo) at the same time as waterworks were constructed in other parts of Japan. The Edo Waterworks[10] was the world's largest facility of its kind, supplying 1.2 million people with water in 1787 [31]. But its facilities all carried spring water or river water, etc. by gravity flow; water stored by dams was still not used.

2.2.1.2   The beginning of modern waterworks

In 1854, the signing of the Treaty of Peace and Amity between the Empire of Japan and the United States started trade with foreign countries and during the following Meiji Period, a modern state was constructed by introducing culture and technologies from Europe and America. On the other hand, cholera, which was then rampant in South-east Asia, arrived in Japan, resulting in severe epidemics in 1822, 1858, 1879, and 1886. Water-borne infectious diseases (cholera, diarrhea and typhoid fever) were contracted by 820,000 people, and 370,000 died as a result between 1877 and 1887 [32].

One response was to build sewage systems and waterworks, mainly in port cities that were in danger of the spread of cholera, as a radical measure to ensure hygienic conditions and fight water-borne diseases. It was also recognized that modern waterworks must be constructed as a measure to prevent fires in the many wooden buildings used at that time.

Under these conditions, in 1890, the principle that waterworks would be run by public enterprises was established, accompanied by the enactment of the national

---

10 Edo waterworks: it was relatively easy to obtain fresh pure water from wells on the plains of Japan, but Edo, which was built on land reclaimed from the inner bay, was not endowed with spring water, so it was dependent on waterworks.

Table 2.2.1 Municipal waterworks projects around the beginning of the 20th century.

Year	Number of project	Name of main City
1889–1892	3	Yokohama-City, Hakodate-City, Nagasaki-City
1893–1897	1	Osaka-City
1898–1902	3	Tokyo, Hiroshima-City, Kobe-City
1903–1907	4	Okayama-City, Shimonoseki-City, Sasebo-City, Akita-City
1908–1911	11	Iwamizawa-City, Yokosuka-City, Aomori-City, Sakai-City, Niigata-City

Source: Japan Water Works Association [33].

government's Waterworks Ordinance. Because Japan lacked any technology or experience in modern waterworks, when their construction began, British waterworks technology was introduced with the technological guidance of many foreign experts, including two British Engineers, Palmer and Barton. Thus began the construction of so-called modern waterworks: "systems that used pressure to provide a continuous supply of clean filtered water through iron pipes."

The first modern waterworks project in Japan was the Yokohama city waterworks of 1887. By 1910, waterworks projects had spread to more than 20 cities, as shown in Table 2.2.1 [33].

When the former Imperial Japanese Navy's naval harbors were constructed in Kure, Sasebo, Maizuru and Ominato during this period, naval port waterworks were constructed to supply water to military facilities and to the warships, laying the foundations for the spread of waterworks to these cities.

In regions short of water resources, where it was difficult to obtain stable supplies, dams introduced to store water were the first modern waterworks constructed in Japan. Modern dams constructed as part of waterworks projects include the Hongochi Upper Dam, the Gohonmatsu Dam and the Sasanagare Dam (CB, 25.3 m, #8), the forerunners of modern dam construction.

These dams played the following roles in ensuring a supply of water in these regions.

2.2.1.2.1 Hongochi Upper Dam

In Nagasaki City on Kyushu Island, because of a shortage of drinking water and the decline in water quality spurred by population growth, cholera claimed the lives of more than 600 people in 1885. In the following year, a Nagasaki Prefecture engineer, Yoshimura Chosaku[11], enacted a waterworks plan in response to a proposal by the

---

11 Yoshimura Chosaku: 1860–1928. After working as an assistant professor at a university, he participated in planning waterworks in the cities of Nagasaki, Osaka, Hiroshima, Kobe and Okayama. He joined the navy, taking part in planning facilities such as the naval port waterworks in Sasebo and also worked as an advisor on waterworks expansion projects in Moji, Kokura, Fukuoka, Sasebo, Nagano and other cities. Overall, Yoshimura made a splendid contribution to the field of waterworks in Japan.

British Engineer, J.W. Hart. The project overcame the movement opposing the construction of the waterworks and financing difficulties, and at the end of March 1891, it was completed as the third modern waterworks in Japan [34]. The relationship between the planned supplied population and quantity supplied by the waterworks expansion project with the municipal water supply capacity ensured by the dam is shown in Figure 2.2.1.

Nagasaki City is surrounded by mountains that are close to the coast and has no large rivers. Because of these topographical conditions, the Hongochi Upper (ER, 28.2 m, #125) and Lower (PG, 26.9 m, #125) Dams, Nishiyama Dam (PG, 40.0 m, #126), and Ogakura Dam (PG, 41.2 m, #122) were constructed at the same time as the earliest waterworks. In recent years, water resources have been ensured by the construction of the Kayaze (PG, 51.0 m, #121), Konoura (PG, 51.0 m, #124) and Yukinoura (PG, 44.0 m, #123) Dams and others, which also functioned as flood control dams [34].

The Hongochi Upper Dam (Figure 2.2.2), which was Japan's first municipal water supply dam, is a zoned earth-fill dam with a height of 18.6 m and effective capacity of 359,000 m³. The waterworks were planned and designed by introducing dam reservoir planning technology that was part of British waterworks planning. This technology was the foundation of many later urban waterworks projects.

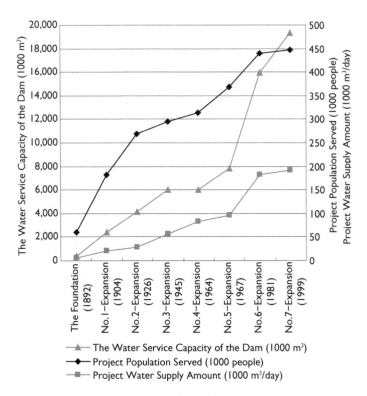

Figure 2.2.1 Changes in waterworks plan for Nagasaki City.

Source: Nagasaki City [35].

Presented by Nagasaki Prefecture

Figure 2.2.2 Hongochi Upper Dam in Nagasaki City.

The facilities were constructed by using many foreign products, such as steel materials (cast iron pipes and valves), cement, etc.

A new concrete gravity dam is now under construction directly upstream from this dam, as an emergency measure taken following the Nagasaki flood disaster of 1982. To preserve the older dams as a cultural legacy, the areas surrounding old and new dams are being improved [36].

2.2.1.2.2 Gohonmatsu Dam

In Kobe City in the Kansai Region, a waterworks project was started in 1892 in response to a severe cholera epidemic that claimed more than 1,000 lives in 1890. At first, the British engineer Barton was in charge, and after the project was suspended during the Sino-Japanese War, Japanese engineers such as Sano Tojiro[12] revised the plan. In 1900, it began to supply water as Japan's seventh modern waterworks system. This waterworks project was first planned to serve 250,000 people by supplying a maximum of 25,000 m³ of water per day. It obtained its water from two dams: the Gohonmatsu Dam, otherwise called Nunobiki Dam, completed in 1900 and the Tachigahata Dam (PG, 33.3 m, #95) that was completed in 1905 and raised in 1915. The first expansion, completed in 1921, provided a maximum daily quantity

---

12 Sano Tojiro: 1867–1929. Tojiro helped Yoshimura Chosaku complete the Gohonmatsu Dam and supervised the expansion of the Sengari Dam. Later, he worked on hydropower dam projects, completing the Oi Dam. He also participated in the construction of the Honen-ike Dam (Kagawa Prefecture) and the Wushanto dam (Taiwan).

Presented by Kobe City [38]

Figure 2.2.3 Gohonmatsu Dam during its construction.

of 104,200 m^3 of water to a supplied population of 500,000, so the Sengari Dam (PG, 42.4 m, #92) was planned as a water resource and completed in 1919 [37].

In this way, Kobe City, which lacked large rivers or lakes to supply water, managed to ensure water for its early waterworks by constructing dams.

The Gohonmatsu Dam is a gravity dam made of rubble concrete[13] with a rubble cement mortar masonry exterior. It was constructed using epoch-making technology of the time: installing pipes inside the dam body to drain seepage, or installing a cutoff wall in the dam foundation (Figure 2.2.3).

It is noteworthy that although the Gohonmatsu Dam is located near the hypocenter of the Kobe Earthquake that caused severe damage in the Kobe-Osaka-Awaji Regions in January 1995, the earthquake did not damage it severely. In addition to dam body reinforcement work for earthquakes that started in 2001, sediment has been removed from the reservoir and the shoreline has been improved.

2.2.2 The spread of municipal water supply projects and the role of dams

2.2.2.1 The spread of waterworks and their effect

Modern waterworks in Japan spread steadily from the early years of the Meiji Period through the twentieth century. During the postwar period of rapid economic growth, the waterworks coverage almost doubled in a single decade (1960 to 1970). In 2001, waterworks supplied water to more than 122 million people (Figure 2.2.4) [39].

---

13 Rubble concrete: concrete with rocks (rubbles) of a diameter from 7.5 to 30 cm.

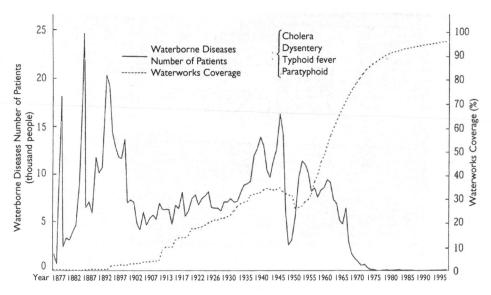

Figure 2.2.4 Waterworks coverage and number of patients contracting water-borne diseases.

Source: Japan Water Works Association [40].

The number of patients suffering from water-borne diseases[14] that prompted the construction of modern waterworks has, as shown in Figure 2.2.4, fallen sharply along with a rise in their coverage since World War II, and the provision of water-works as a hygiene measure has achieved its purpose.

2.2.2.2   Ensuring water resources and dam construction

2.2.2.2.1   Ensuring water resources and the legal system established by the national government

The sources of water were often groundwater and surface water from rivers (natural flow) during the early municipal water project era, so obtaining stable water resources was a major concern for waterworks managers. Municipal water projects were, in principle, publicly operated by municipalities under national government laws, so local authorities made every effort to obtain its own water resources as necessary. During the postwar rehabilitation period and into the rapid economic growth period, the concentration of the population in cities, improvements in life styles and the re-markable increase in demand for industrial water rapidly expanded all uses of water supplied by waterworks.

As the development and use of water resources advanced to a new level in this way, the Waterworks Law was enacted by the national government in 1957 in order to

---

14 Water-borne diseases: diseases that spread through a water medium. Typical diseases of this kind are cholera, typhoid fever, diarrhea and hepatitis A.

develop organizations to construct and manage waterworks. To rationally develop water resources and to effectively use the water resources developed, while harmonizing flood control and water use and coordinating water uses, the Specified Multi-purpose Dam Law (1957), the Water Resources Development Promotion Law and the Water Resources Development Public Corporation Law (of 1961, abolished in 2002, and the Japan Water Agency Law was promulgated in the same year), and other laws concerning water resources development have been enacted. Along with the promulgation of these laws, the development of water resources for waterworks in large cities and elsewhere moved ahead to the age of construction of multi-purpose dams.

2.2.2.2.2 The number of dams for waterworks and the capacity of waterworks

Table 2.2.2 shows the number of dams constructed for waterworks by decade. Since the 1960s, the number of dams has risen, mainly because of the construction of multi-purpose dams by Comprehensive River Development Projects. In 2002, more than 180 dams for waterworks were either being constructed, or were planned.

Figure 2.2.5 shows changes in the population served by waterworks and the capacity of water supply dams. The construction of dams has expanded the capacity for waterworks, supporting the rise in the supplied population. By 1995, it had reached 1.26 billion m^3.

2.2.2.2.3 Amount of water obtained by dams

Next, the roles of dams that ensure water resources are examined from the perspective of amount of water obtained by constructing dams. Table 2.2.3 shows the amount of water, by sources, applied to supply waterworks systems in Japan. The total amount of water in 1999 was 16.87 billion m^3, an increase of about 2.6 times the total amount of water of 6.45 billion m^3 in 1965. The amount of water ensured by constructing dams during this period accounts for 55% of the total increase,

Table 2.2.2 Number of dams constructed for waterworks, in chronological order.

Year	The number of dams for waterworks only	The number of dams which are concerned with waterworks	Total
1990–1925	16	1	17
1926–1945	16	6	22
1946–1955	8	9	17
1956–1965	7	23	30
1966–1975	21	63	83
1976–1985	27	81	108
1986–	8	382	390
Total	103	565	668

Source: Japan Water Works Association [41].

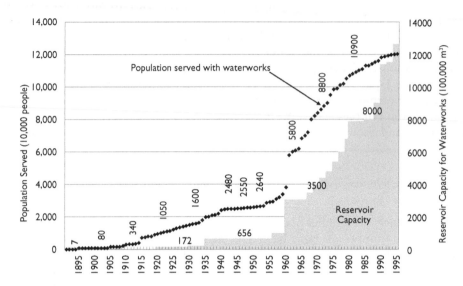

Figure 2.2.5 Population served by waterworks and reservoir capacity for waterworks.

Source: MLIT [42].

Table 2.2.3 Amount of water supplied to waterworks by water sources.

Year	River	Dam	Lake water	Underflow water	Well	Others	Total
1965	35.8	7.5	0.6	9.5	9.4	1.7	64.5
	55.5	11.6	0.9	14.8	14.6	2.6	–
1975	58.0	27.8	1.2	10.7	24.9	2.8	125.4
	46.3	22.2	0.9	8.5	19.9	2.2	–
1985	52.6	48.9	1.9	7.4	33.1	4.9	148.8
	35.3	32.9	1.3	5.0	22.2	3.3	–
1999	53.6	65.0	2.3	6.5	36.2	5.1	168.7
	31.8	38.5	1.4	3.8	21.5	3.0	–

Note: The lower shows the percentage.

Source: Ministry of Health, Labour and Welfare [43].

from 0.75 billion m³ to 6.5 billion m³. The percentage of the total amount of water provided by dams, among all sources, which was about 12% in 1965, had risen to about 39% in 1999.

In this way, the roles that dam projects have played in ensuring stable water resources for waterworks increased through the process of the spread of modern waterworks. Since 1965 in particular, rapid growth has been accompanied by an improvement in people's lifestyles and it is no exaggeration to state that dams have ensured the supply of water from the adjacent prefectures widely.

2.2.3 Examples of roles played by dams in municipal water supply projects

2.2.3.1 Roles of waterworks projects and dams in metropolitan Tokyo

The Ogochi Dam (PG, 149.0 m, #49) was constructed to ensure water resources for metropolitan Tokyo during the postwar rehabilitation period. It is one of the world's largest dams specialized for waterworks (effective reservoir capacity: 185.4 million m³), was completed in 1957, and still functions as the metropolitan region's water source. When the Ogochi Dam was completed, a commemorative stamp was issued, showing how great the expectations of dams were under social conditions characteristic of the postwar rehabilitation period (Figure 2.2.6).

Metropolitan Tokyo, where the Ogochi Dam was constructed, started waterworks service in 1898, when a maximum of 167,000 m³/day were supplied to about 1.5 million people. Expansion projects began during the Taisho period (1912–1926) to respond to the concentration of population in Metropolitan Tokyo and as waterworks expansion on the Tama and Tone River Systems created new water resources, a series of expansion projects achieved the present supply system that provides 6.96 million m³ of water per day to the 11 million residents of the metropolis [44].

Shortly after waterworks were first introduced in Tokyo, demand during the summer season exceeded the supply capacity by 50%. So as part of the First Extension Project of Waterworks (1912), the Murayama (TE, 32.6 m, #47) and the Yamaguchi (TE, 35.0 m, #44) Reservoirs were constructed, and as part of the Second Waterworks Expansion Project (1936), construction of the Ogochi Dam, which obtains water from the Tama River that runs across Tokyo, was planned. The construction of the Ogochi Dam was temporarily suspended because of a shortage of materials and labor during World War II, but resumed after the war and was completed in 1957 [45, 46].

Figure 2.2.6 Commemorative stamp of the Ogochi Dam.

Later, because of an increase in water demand caused by a concentration of population, an unprecedented water shortage in 1964, the year of the Tokyo Olympics, caused severe social problems. As a result, as stated in section 2.5, starting in 1965, water resources were obtained by constructing multi-purpose dams such as the Yagisawa (VA, 131.0 m, #31), Shimokubo (PG, 129.0 m, #40), Kusaki (PG, 140.0 m, #35) and Naramata (ER, 158.0 m, #32) Dams [47]. Now, as shown by Figure 2.2.7, approximately 70% of water for waterworks in Metropolitan Tokyo is provided by a group of dams on the Tone River, situated outside Metropolitan Tokyo. In the Figure 2.2.7, the population became concentrated in Tokyo from 1920 to 1940, rapidly increasing the supplied population, but this fell sharply at the end of World War II.

The Ogochi Dam played a major role as a waterworks supply source during the first half of the 1960s when waterworks in Metropolitan Tokyo were dependent on the Tama River System. The waterworks networks of the Tone River and Tama River Systems are now linked, so that reservoirs on the Tama River System are used during the high demand summer period and during accidents or droughts on the Tone or Ara Rivers.

In the catchment basin of the Ogochi Dam, the Tokyo Waterworks Bureau that administers waterworks manages water resource forests intended to nurture the river

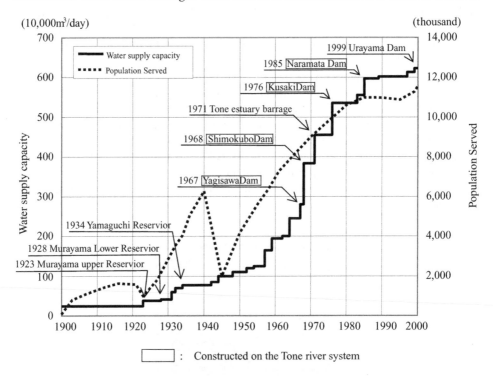

Figure 2.2.7 Expansion of Tokyo Waterworks – change in the population served and water supply capacity in Tokyo Metropolis.

Source: Tokyo Metropolitan Government [48].

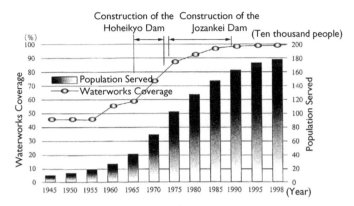

Figure 2.2.8 Served population and waterworks coverage and dam construction in Sapporo City, Hokkaido

Source: T. Ida [49].

head for municipal water system in order to ensure stable river flow and conserve water resources in the Tama River Water Source Region.

2.2.3.2   Roles of waterworks projects and dams in Sapporo City

As shown by Figure 2.2.8, the construction of two dams, the Hoheikyo Dam (PG, 102.5 m, #3) and the Jozankei Dam (PG, 117.5 m, #2) play a major role in increasing the waterworks coverage and the served population in Sapporo City.

The effectiveness of the dams constructed by these projects is shown by a simulation of the water-shortage rate and number of water shortage days when water is not supplied by the dams, using the flow rate over a 38-year period (1960 to 1997). It is reported that, according to the results, without dams, water saving measures would be necessary every year, the water shortage rate would be an average from 20% to 45% and a maximum of 35% to 70%, and the number of water-shortage days would average 130 days [49].

As this shows, dam construction permits stable water use and prevents suspension of the water supply.

2.3   HYDROPOWER DAMS THAT HAVE SUPPORTED THE GROWTH OF MODERN INDUSTRY

With the appearance of the first practical working electric generators in Germany around 1870, industry began using electricity. At the same time, hydropower production from generators driven by water turbines began, adding hydropower production to the roles of dams.

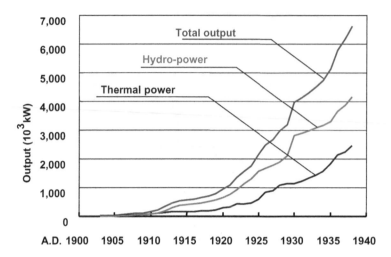

Figure 2.3.1 Changes of electric power production at the early stage.

Source: Electric Power Civil Engineering Association [51].

Dams that had provided a stable supply of water for irrigation and waterworks, were given a new unique role in creating the head needed for hydropower production. Hydropower has contributed to the growth of industry.

2.3.1 Hydropower production and dams

Hydropower production that began with waterwheels on small rivers has expanded to include the run-of-river type, conduit type, dam and conduit type, and dam type.

The important role of dams and reservoirs with stable intake and reservoir functions in obtaining the potential energy of water is fulfilled by the hydropower production process.

During the last half of the 1880s in Japan, hydropower production appeared as an economical power production method to replace coal-fired thermal power production, meeting the growing electric demand. The first hydropower was run-of-river type with small-scale intake weirs installed to stabilize the intake water level. In about 1900, the construction of large-scale hydropower plants in mountainous regions far from demand regions began in response to progress in long-distance electric power transmission technology. From about 1910, the hydro first – thermal second[15] stage arrived, and the construction of hydroelectric stations as part of dam regulation pond construction began. In the 1920s, dam-conduit type hydropower plants appeared,

---

15 Hydro first – thermal second: a phrase that indicates that hydropower plants provide more power than thermal electric power plants.

providing a base load supply[16] in response to soaring demand for industrial electric power [50].

This section describes hydropower and the role of dams from about 1890, when hydropower began in Japan, to 1945 when full-scale generation of hydropower by dams started.

Figure 2.3.1 shows changes of electric power production during this period.

2.3.2   The start of hydropower production

Hydropower production was first developed for in-house use by the spinning and mining industries. The first electric power plant developed to provide commercial electric power was constructed in Kyoto: the Keage Power Plant (1892) that used water drained from Lake Biwa (conduit type). Its power was used to operate the first electric street cars in Japan. The Lake Biwa Canal project, planned under the leadership of Tanabe Sakuro[17], was undertaken to stimulate industry in Kyoto, which had declined since the capital was moved from Kyoto to Tokyo in 1869. The purpose of this project was to construct a shipping canal linking Lake Biwa with the Uji River in Kyoto by cutting a canal to Lake Biwa with its rich water resources and at the same time using water from Lake Biwa to generate hydropower, irrigate farm fields, and fight fires.

Demand for electric power for lighting began in 1887 and records of electric power demand for factories appeared in 1903, when Japanese industry finally modernized.

Early electric power projects were primarily intended to supply electric power for lighting from thermal power plants. During this period, transportation within Japan was inconvenient and transporting coal was costly, so it was difficult to produce thermal power in inland regions of Japan. Therefore, most power produced in such regions was hydropower. In other words, hydropower development began in regional cities close to hydropower zones.

Many water intake systems used at hydropower plants at that time were made by packing boulders obtained on the scene into frames of assembled logs. As an example, Figure 2.3.2 shows a weir at the Koyama Power Plant (Chubu Electric Power Co., 1910) in the midstream area of the Oi River in the Chubu Region. Table 2.3.1 is a Table of the oldest hydropower plants in various regions.

---

16 Base load supply: the supply capacities of power sources are generally categorized into three types: a) base load supply capacity that operates constantly at an almost fixed output, b) peak load supply capacity that operates in response to electric power demand fluctuations to supply power, mainly when is needed at peak demand times, and c) middle load supply capacity that plays an intermediate role between the first two. Base load supply capacity is used relatively often, so as a power source that provides both superior long-term economic benefits and stable fuel procurement, it is supplied by nuclear, coal-fired thermal, run-of-river hydroelectric and geothermal power plants.

17 Tanabe Sakuro: 1861 to 1944. After graduating from the Faculty of Civil Engineering at the Imperial College of Engineering in 1883, he went to work for Kyoto Prefecture. He was a pioneer in civil engineering in Japan during his active years, from the late 19th century to the 1920s. He was the senior engineer in charge of the design and execution of the Lake Biwa Canal Project, and he also constructed the accompanying Keage Power Plant in 1890, establishing Japan's first commercial hydropower plant.

Presented by CEPCO

Figure 2.3.2 A weir at the Koyama Power Plant.

Table 2.3.1 Oldest hydropower plants in each region.

Region	Name of power plant	River system	Effective head (m)	Maximum discharge (m³/s)	Maximum output (kW)	Beginning of opera-tion	Classifi-cation	Current state
Tohoku	Sankyozawa	Natori	26.67	5.57	5	1888.7	In-house use	1000 kW operating
Kanto	Shimotsuke Asa Bouseki (Owner)	Tone	–	–	17	1890.7	In-house use	Abolition
Chubu	Iwazu	Yahagi	53.94	0.37	50	1897.7	Project use	130 kW operating
Kansai	Keage	Yodo	33.74	16.7	80 × 2	1891.11	Project use	4500 kW operating

Source: Electric Power Civil Engineering Association [52].

2.3.3 The start of long-distance transmission of electric power and large hydropower dams

After the Russo-Japan war, the Japanese economy underwent rapid growth. Because electric power demand was also expanded rapidly by the Russo-Japan war, the electric power industry acquired an important position in Japanese industry. This growth of electric power demand grew in two areas: spreading electric lighting in homes and the electrification of power provision in factories.

The spread of electric lighting helped young people study at night, improving literacy and contributing to increasing the strength of the nation. At that time, young people were a valuable source of labor with many working during the day, so the spread of electric lighting provided them with an opportunity to study at night.

The earliest hydropower plants in Japan were extremely close to their demand regions, and their generator output and transmission voltage were both low. However, in 1899, the transmission of 11 kV for 26 km and the transmission of 11 kV for 22 km were achieved in the Chugoku and Tohoku Regions, respectively, permitting longer distances between hydropower plants and consumption regions, thereby contributing greatly to electric power production projects in Japan. Later, electric power companies worked to increase transmission voltages, lengthen transmission distances and to develop high-capacity hydropower plants.

During this period, intake facilities used to generate electric power also changed as low fixed water intake weirs that could take in the flow rate in the dry season were replaced by dams with gates, and these were expanded to include dams with regulating ponds. Large scale hydropower plants developed in this way are shown in Table 2.3.2.

Of these, the Shimotaki Power Plant in the northern Kanto Region supplied power to Tokyo at that time, providing almost the entire demand (approx. 40 million–80 million kWh/year) to run trams in Tokyo. The Kurobe Dam (PG, 33.9 m, #26), constructed as the water intake dam for the Shimotaki Power Plant, which is Japan's first concrete gravity dam for hydropower, has a total reservoir capacity of 2.366 million m^3 (effective reservoir capacity: 1.160 million m^3).

In addition, the Yatsuzawa Power Plant (Tokyo Electric Power Company, Inc. (TEPCO), 1912) in western Kanto was not only a high-capacity dam, but also a conduit type with a large regulating pond (effective capacity: 467,000 m^3). It was an epoch-making type of dam at that time. The Ono Dam (TE, 37.3 m, #62, Figure 2.3.3), which formed this large regulating pond was the largest earth dam in Japan at that time.

The development of large-scale hydropower plants had a number of important impacts on the management of the electric power industry. First it allowed a drop in the price of electricity, because hydropower could be produced more cheaply than thermal power. Secondly, it permitted companies to meet the daytime demand for power for industry, in addition to the nighttime demand for lighting power.

Because most hydropower plants were the conduit type at that time, it was impossible to control daytime and nighttime flow rates. This means that when an appropriate customer could not be found, it was impossible to produce power using

Table 2.3.2 Large scale hydropower plants constructed by the beginning of 20th century.

Name of power plant	Name of river system	Dam or Water resource	Beginning of operation	Maximum output (kW)	Voltage (V)	Distance (km)
Komabashi	Sagami	Lake Yamanaka	1907.12	15,000	55,000	75
Yaotsu	Kiso	Lake Maruyama Sosui	1911.12	7,500	66,000	34
Yatsuzawa	Sagami	Ono dam	1912.7	35,000	55,000	75
Shimotaki	Tone	Kurobe dam	1912.12	31,000	66,000	125
Inawashiro No.1	Agano	Lake Inawashiro intake weir	1914.10	37,500	115,000	225

Source: Electric Power Civil Engineering Association [52].

Presented by TEPCO

Figure 2.3.3 Ono Dam of the Yatsuzawa Power Plant. (See colour plate section)

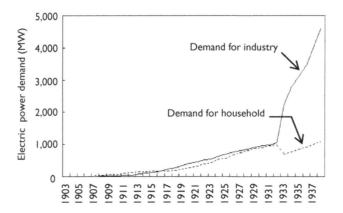

Figure 2.3.4 Transition of maximum electric power demand before World War II.
Source: Electric Power Civil Engineering Association [53].

the daytime flow rate when demand for electric power was lower than at night, and this encouraged a rise in electric power production costs. Electric power companies attempted to obtain daytime demand by lowering their daytime electricity charges, contributing to the profitability of industries that used this cheap electric power (Figure 2.3.4).

During this phase, the structure of electric power production facilities underwent a sharp change, from thermal first–hydro second to hydro first–thermal second. Hydro surpassed thermal power in 1911, ushering in the age of hydro first–thermal second in the Japanese electric power industry: for about a half century from 1911 to 1960 [54].

2.3.4   The development of dams and conduit type high-capacity hydropower production

During the Taisho Period (1912–1926), the Japanese economy was affected by world-wide economic growth, resulting in lively growth of about 5% per annum of Japan's manufacturing industries, until the start of World War II. The growth of the electro-chemical industries and the machinery and iron and steel sectors after World War I was remarkable. To support production, these industries required an abundant and low-priced supply of electricity. The electric power supply grew explosively at a rate in excess of 20% per annum, and the first shortage of electric power since the establishment of the industry occurred during a drought in 1918.

Other reasons for the rapid development of hydropower were the successful introduction of long-distance power transmission, permitting the development of large hydropower production in mountainous regions, and the flourishing of industries that used low-priced electric power available at times when electric power demand was low. Operating thermal power plants in parallel to supplement hydropower production during dry seasons ensured stable electric power and encouraged the expansion of the electric power industry. This gave the industry the idea of effectively using a quantity of water in excess of the flow rate in the dry season by creating complementary hydropower–thermal power systems, and increased the maximum intake to approximately the average water flow rate.

Under these circumstances, companies that planned and conducted large-scale hydropower development were established, one after another. They developed the rich hydropower of mountainous areas in the Chubu Region and constructed high-voltage transmission lines to supply electric power to cities. One example was the successful transmission of 154 kV for 238 km, from the Suhara Power Plant (Kansai Electric Power Co., Inc. (KEPCO), 1922) in Chubu to the Osaka Substation in Osaka (now the Furukawabashi Substation), in 1923. The achievement and spread of long-distance electric power transmission, by increasing voltage, spurred hydropower development in mountainous regions, with particularly remarkable development of high capacity hydropower beginning in the late Taisho period (1912–1926).

An example is the Oi Electric Power Plant that includes the Oi Dam (PG, 53.4 m, #70) in Chubu Region. The Oi Electric Power Plant, a dam – conduit type power plant developed on the Kiso River, was completed in 1924. It was originally planned as a conduit type, but it was converted to a dam type that can respond to peak demand, making it the first power plant to include a large-scale dam constructed in Japan. As a concrete-gravity dam with a height of 53.4 m, it was the highest dam in Japan at that time. The maximum output of this power plant was equivalent to half of the entire electric power demand in Aichi Prefecture at that time.

The Shizugawa Power Plant (KEPCO), which includes the Shizugawa Dam in Kyoto Prefecture, developed in 1924, provided more than 10% of all electric power used in Osaka Prefecture at that time. When it was developed, the Osaka – Kyoto – Kobe Metropolitan Region was particularly short of electric power, so its development made a big contribution to the supply of electric power in the region.

Table 2.3.3 shows typical dam type and dam-conduit type hydropower plants that were constructed in various districts during the Taisho Period.

Table 2.3.3 Hydropower plants representing each region after World War I.

Present owner	Name of power plant	Name of river system	Name of dam	Height of dam (m)	Maximum output (kW)	Beginning of operation
HEPCO	Nokanan	Ishikari	Nokanan (PG, 30.0 m, #1)	30	5,100	1918
CEPCO	Kamiasou	Kiso	Kamiasou Hosobidani (PG, 22.4 m, #68)	22.4	24,300	1926
KEPCO	Shizugawa	Yodo	Shizugawa	35.2	32,000	1924
KEPCO	Oi	Kiso	Oi	53.4	42,900	1924
ENERGIA	Taishakugawa	Takahashi	Taishakugawa	62.1	3,706	1924

Source: Electric Power Civil Engineering Association [52].

2.3.5   The increased use of river water as an energy source

During the early Showa Period (1926–1945), large capacity hydropower development continued in response to the results of economic evaluations of two approaches that began to spread in the late Taisho Period: making the maximum intake quantity approximately the average water flow rate and using thermal electric power as supplementary power during dry seasons. At the same time as national government control of industry strengthened, mining and manufacturing industry production soared, and metal, chemical, and machinery industries grew at a particularly rapid rate. Electric power companies responded to trends in the manufacturing industry by devising and implementing the concept of successively developing hydropower plants mainly from the downstream reaches of large-scale rivers.

Of these, the development of hydropower on the Kurobe River in Hokuriku Region began with the completion of the Yanagawara Power Plant (1927), and moving upstream, was followed by the Kurobegawa No. 2 Power Plant (1936) supplied by the Koyadaira Dam (PG, 54.5 m, #56), then the Kurobegawa No. 3 Power Plant (1940) supplied by the Sennindani Dam (PG, 43.5 m, #57). At the same time, the Aimoto Power Plant (Toyama Prefecture, 1936) was completed downstream from Yanagawara.

Among these, the Kurobegawa No. 3 Power Plant was a historical project that had to overcome problems such as avalanches and high-temperature tunnel work, etc. to be completed at the early stage of World War II.

When the national government took control of electric power, continued surveys moved upstream, but because it was followed shortly by World War II, hydropower development ended with the construction of the Kuronagi No. 2 Power Plant (1947) on a tributary. The later successive development of the Kurobe River is described in 2.5.3 (2) (b).

As successive developments were carried out along the river, dams used exclusively to produce hydropower were constructed separately at locations in the river basin where topographical conditions suited hydropower development, advancing the use of river water as an energy source.

In the 1920s, on the Oi River System in Chubu Region, the water intake dam, the Tashiro Dam (PG, 17.3 m, #76), was constructed as the furthest upstream dam

Presented by CEPCO

Figure 2.3.5 Yasuoka Dam. (See colour plate section)

Table 2.3.4 Hydropower plants representing each region before World War II.

Present owner	Name of power plant	Name of river system	Name of dam	Height of dam (m)	Maximum output (kW)	Beginning of operation
KEPCO	Komaki	Sho	Komaki	79.2	72,000	1930
KEPCO	Kurobegawa No. 3	Kurobe	Sennindani	43.5	81,000	1940
TEPCO	Tashirogawa No. 2	Oi	Tashiro	17.3	20,362	1928
CEPCO	Yasuoka	Tenryu	Yasuoka	50.0	52,500	1936
CEPCO	Oigawa	Oi	Oigawa (PG, 33.5 m, #79)	33.5		
			Sumatagawa (PG, 34.8 m, #80)	34.8	62,200	1936

Source: Electric Power Civil Engineering Association [52].

located 160 km from the river mouth, and hydropower plants (Tashiro River No. 1 and No. 2 Power Plants) were developed, carrying the water into the Hayakawa River on the Fuji River System. This hydropower was transmitted to the Metropolitan Tokyo area. The power produced by these power plants was equivalent to about three times the demand by Tokyo at that time.

In the middle reaches of the Oi River and on the Tenryu River, dam type hydropower plants were constructed, forming the core electric power development of each river system at that time. The maximum output of the Oigawa Power Plant was so massive that it equaled approximately half of the contract kilowatts for all electric power in Shizuoka Prefecture at that time. The Yasuoka Dam (PG, 50.0 m, #67, Figure 2.3.5), the first dam constructed on the Tenryu River to produce electric power, was also completed during that period.

Table 2.3.4 shows representative dam type and dam-conduit type hydropower plants that were constructed in various regions during that period.

2.4   POSTWAR REHABILITATION AND MULTI-PURPOSE DAMS – THE APPEARANCE OF COMPREHENSIVE RIVER DEVELOPMENT DAMS

Multi-purpose dams that also control floods were first seen in Japan around 1950. This concept appeared in the 1920s. Multi-purpose dams supported the rehabilitation of Japan following World War II with flood control, irrigation and electric power production, under a policy of Comprehensive National Land Development.

2.4.1   The dawn of multi-purpose dams with flood control

Dam projects that were intended mainly to control floods first appeared in river improvement project plans prepared in 1922: the Kinu River Improvement Plan[55] was enacted in the following year. The dam that was planned at this time was realized in later years, as the Ikari Dam (PG, 112.0 m, #25). In 1927, Aoyama Akira[18], who had participated in the construction of the Panama Canal, took part in the geological survey as a branch manager.

At about this time, engineers of the Ministry of the Interior, who had observed projects in Europe, began to advocate River Flow Regulation Projects, planned by constructing multi-purpose reservoirs with electric power, irrigation and other water uses added to flood control, thereby unifying flood control and water usage. Dr. Mononobe Nagaho[19], who planned the Kinu River Improvement Project, offered the following proposals in a paper published in 1926 [56]. The paper's main points were:

a   In Japan, the period for which a river course displays its full capacity is extremely short, so controlling the river flow rate with reservoirs is an extremely effective way to prevent floods.

b   Electric power production is hampered by droughts in winter, and during this season, there is no danger of a large flood, so the flood control capacity can be used to generate electric power.

18 Aoyama Akira: 1880–1963. Aoyama participated in the Panama Canal Project reopened in 1904. After returning to Japan, he completed the Arakawa Diversion Channel, which was constructed to protect Tokyo from flooding. Then, to protect the Niigata Plain from flooding, as the Director of the Ministry of the Interior's Niigata Branch Office, he completed the repair of the Okozu Diversion Channel on the Shinano River, which had ceased to function because of the settlement of its foundations. In his honor, a plaque engraved with the words in Esperanto, "Joyful is the man who discerns the purpose of the Gods in all things. For humanity, for the nation." has been erected at Okozu. In 1934 and 1935 he was Engineer-General of the ministry. In 1934, he prepared two documents: "Creed of a Civil Engineer" and "Guide to the Work of Civil Engineers" for the Japanese Society of Civil Engineers.

19 Mononobe Nagaho: 1888–1941. After joining the Ministry of the Interior, Mononobe served as Director of the Civil Engineering Laboratory and was a professor at the University of Tokyo. One of the pioneers of hydraulics and earthquake engineering in Japan, he provided advanced guidance on river structure design and earthquake resistance technologies. He established the concrete gravity dam design method. He also advocated the river flow regulation theory, establishing integrated river development projects that united entire river systems and multipurpose dams that are a core element of such projects.

c   Reservoir locations should be used for multiple purposes because, in Japan, there are generally few effective locations.

In addition, he stated, "I believe that we must quickly survey the results of flood control by regulating water quantities in rivers, lakes and marshes, and the expansion of various kinds of water usage in Japan, establish related overall policies and to the degree that circumstances allow, establish and implement policies for flood control and water usage projects that should be implemented in the future."

To realize these proposals, the Ministry of the Interior started systematic surveys of River Flow Regulation Projects in some river systems in 1934. Prior to this, work on the Kamaguchi Sluice Gate had begun as partial Tenryu River upstream improvement work, a prefectural project undertaken by Nagano Prefecture in 1932 in order to control the quantity of water in Lake Suwa that had been surveyed, by focusing on flood control problems in the past. Work on the Edo River Flow Regulation Project researched by the Ministry of the Interior centered on municipal water supply problems, resulted in work on the Edo River Lock under a contract by the Minister of the Interior with costs borne as a Tokyo Prefecture waterworks project in 1936. Table 2.4.1 shows the initial River Flow Regulation Project.

Under these circumstances, an investigation budget for the River Flow Regulation Project was established and the Investigation Council for the River Flow Regulation Project was formed within the government in 1937, with the cooperation of the ministries concerned with river water–the Ministry of the Interior, the Ministry of Posts and Telecommunications, and the Ministry of Forestry and Agriculture–initiating the investigation of rivers under government management. This survey was carried out on 64 river systems, including the Tone River, and the target for the entire plan was stipulated as 130 river systems.

Table 2.4.1 Initial River Flow Regulation Projects [57].

Name of river	Project manager	Project cost (10^3 yen)	Cost sharing	Purpose*	Construction period	Name of main dam
Lake Suwa	Nagano Pref.	1,250	Gov. & Pref.	C, H, I	1932–1936	Kamaguchi Sluice Gate (to control discharge from the Lake Suwa)
Edo	Tokyo Pref.	5,137	Pref.	I, S	1935–1938	
Aseishi	Aomori Pref.	2,364	Gov. Pref. & Electric co.	C, H, I	1934–1945	Okiura dam (PG, 40.0 m, #10)
Nishiki	Yamaguchi Pref.	6,200	Pref.	H, S	1938–1940	Kodo dam (H = 43.3 m)
Oirase	Aomori Pref.	32,281	Electric co.	C, H, I	1938–1943	
Sagami	Kanagawa Pref.	150,000	Pref.	H, S	1938–1947	Sagami dam (PG, 58.4 m, #51)
Omaru	Miyazaki Pref.	1,770,000	Gov. & Pref.	C, H	1939–1951	Matsuo dam (PG, 68.0 m, #130)
Tama	Akita Pref.	27,243	Electric co.	H, I	1939–1942	

Note: *C: Flood Control, I: Irrigation, H: Hydroelectric, S: Water Supply.

Source: MLIT.

Many River Flow Regulation Projects were implemented, beginning in about 1938. River Flow Regulation Projects on rivers managed by the national government were expanded to rivers that needed to be added to river improvement plans in response to large floods in 1935 and 1938, and to river flow regulating dams built mainly to control floods on a number of rivers, including the Kitakami River.

These River Flow Regulation Projects were almost all suspended or terminated as World War II intensified, but immediately after the war, became the foundations of the rehabilitation system.

2.4.2   Dams in postwar rehabilitation and in Comprehensive Land Development Plans

2.4.2.1   Continuing large flood disasters and execution of Comprehensive River Development Projects

Immediately after World War II, Japan suffered from shortages of food and energy. Furthermore, it was struck by a series of large typhoons (1945, 1947, 1948, etc., Figure 2.4.1) that inflicted severe losses on the land and the people of Japan, who had still not recovered from the suffering brought about by their defeat.

To resolve these problems quickly, it was necessary to construct so-called multi-purpose dams. These were intended to contribute to comprehensive development by supporting flood control measures based on: regulating flood discharge; ensuring water for irrigation and industry; developing inland forests; and ensuring electric power from reservoir-type electric power plants, thus spurring the construction of reservoir-type hydropower plants at these multi-purpose dams. Construction of major multi-purpose dams included the Yanase dam (PG, 55.5 m, #112, started in 1948, completed in 1954) and the Sarutani Dam (PG, 74.0 m, #99, started in 1950 and completed in 1957) under a River Flow Regulation Project of the national government. Work resumed on dams partially-constructed before being suspended by the war. Two of these were the Tase Dam (PG, 81.5 m, #15) and the Ikari Dam (Figure 2.4.2). River Flow Regulation Projects were renamed Comprehensive River Development Projects[20] in 1951.

Then in 1950, the National Comprehensive Land Development Act was enacted to conserve national land, increase food production, and develop hydropower, and based on this law, Comprehensive Development Plans for Specified Areas, such as the Kitakami River Comprehensive Development Plan for Specified Areas, were enacted.

---

20 Comprehensive River Development Project: a project implemented in accordance with a comprehensive plan that includes flood control and water usage. In other words, to ensure that a river comprehensively fulfils its river management and water usage functions, such a project combines river management purposes that are centered on using dams to control flood discharge, to ensure the flow rate necessary to use the river's water appropriately and to maintain the normal functions of the river flow, with water usage purposes that include powering hydropower plants with river water and developing water resources to supply agricultural and municipal water, and so on.

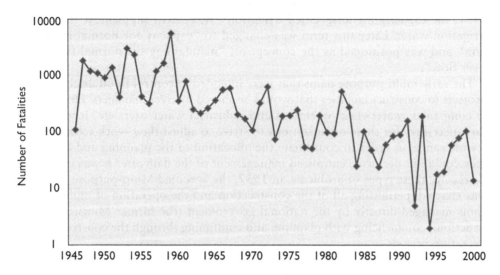

Figure 2.4.1 Fatalities caused by storms and flood disasters [58].

Note: 1. This graph shows a total death toll from natural disasters (floods, landslides and volcanoes).

2. The figures refer to statistics published by the Ministry of land. Infrastructure and Transportation (during 1946–1952) and Tokyo Metropolitan Police Department (from 1953).

Presented by MLIT

Figure 2.4.2 Ikari Dam. (See colour plate section)

For the Fujiwara Dam (PG, 95.0 m, #34), the construction of which began in 1951, the capacity for unspecified irrigation[21] was set to supplement the existing irrigation water. Later this term was changed to "capacity for normal river function" and was positioned as the concept of, "maintaining the normal function of river flow".

The early multi-purpose dams that were intended to control flood discharge were projects to construct facilities that would be shared by river managers, electric power companies, waterworks operators and industrial water users etc. It was, therefore, necessary for these organizations to strive to adjust their work schedules over a wide range, in order to coordinate the allocation of the planning and costs. And they could not perform centralized management of the dam after it was completed. To resolve these types of problems, in 1957, the Specified Multi-purpose Dam Law was enacted, permitting all of the construction and the operation of multi-purpose dams managed directly by the national government (the former Ministry of Construction), commencing with planning and continuing through the construction and operation periods.

2.4.2.2   Electric power shortages and the reorganization of electric power, and the development of large-scale reservoir-type hydropower plants [59]

2.4.2.2.1   Electric power shortages and the promotion of electric power resource development

The end of World War II was followed by a temporary surplus of electric power, because electric power consumption was halved from its former level by stagnation of manufacturing activities caused by the wartime destruction of manufacturing plants. As a consequence of the spread of electric heaters to heat people's homes in response to shortages and soaring prices of coal, petroleum and gas, and of the spreading use of electric power that could be obtained easily and cheaply as a power source to restore manufacturing, electric power demand soared. Annual energy supply that was down to 19.5 billion kWh in 1945 had leaped to 29.4 billion kWh in 1947.

However, new electric power source were not developed, as little work was done to restore electric power systems damaged by the war and to continue projects initiated before the war. Later, at the end of 1949, approval was given for hydropower development at 33 locations, with an intended production of 1,180 MW, as hydropower development funded by the US Economic Rehabilitation Fund.

The system of state control of the electric industry that had been implemented through Japan Power Generation and Transmission Co. Inc. during the war, ended with the 1951 breakup of the electric power industry into the current nine regional companies as an occupation policy of moving away from Japan's over-centralized economy. That year, the Korean War that boosted electric power demand was accompanied by

---

21 Unspecified irrigation: This term means the irrigation that is performed for unspecified existing farm land. In contrast to this, "specified irrigation" is defined as irrigation performed for a specified region or for the benefit of certain specified water users.

an extremely severe drought in the autumn, resulting in an unprecedented electric power crisis. At that time, frequent power failures made candles a standard form of emergency lighting in homes.

Such circumstances triggered demand for the immediate start of work to develop large-scale hydroelectric source. In 1952, the Electric Power Development Promotion Act was enacted. Under this law, the Electric Power Development Co., Ltd. (J-POWER) was founded with government funding, to establish a power source development system with the primary task of directly investing government funds in regions where development was difficult. This law also stipulated that the Electric Power Development Coordination Council would prepare long-term basic plans for electric power and annual implementation plans, including all electric power development projects conducted by electric power companies and public bodies. In these ways, the postwar electric power development system was established.

2.4.2.2.2   Development of large-scale dam type hydropower plants

When the growth of hydropower production in Japan began, it was centered on run-of-river type hydropower plants that required relatively little initial funding and until the 1950s, the decline in hydropower production during the drought season was supplemented by thermal power production. In the late 1950s, of the hydropower plants at approximately 1,460 locations, only about 40 were equipped with reservoirs that could regulate their flow.

Thermal power plants operating after the war were powered by coal, but their electric power production efficiency and profitability both fell remarkably because of delayed supplies of coal, a decline in its quality, and a rise in its price. As a result, the construction of dam type hydropower plants was re-emphasized as a way of increasing water usage and overcome the seasonal imbalance.

More advanced thermal power plants were being constructed to provide electricity to meet rising demand in response to the postwar rehabilitation of industry, and with these stations providing base load, large dam type hydropower plants that were intended to meet the peak demand for electric power, increased in importance, spurring their construction. Table 2.4.2 shows- the major hydropower dams that were completed during this period.

Table 2.4.2 Large-scale hydropower dams in the 1950s.

Name of Dam	Owner	River	Dam height (m)	Type	Name of Power Plant	Output (MW)	Commencement of Operation
Maruyama (PG, 98.2 m, #71)	KEPCO	Kiso	96	PG	Maruyama	125	1955
Kamishiiba (VA, 110.0 m, #129)	Kyusyu EPCO	Mimi	110	VA	Kamishiiba	90	1955
Sakuma	J-POWER	Tenryu	155.5	PG	Sakuma	350	1956
Ikawa (Figure 2.4.3)	CEPCO	Oi	103.6	HG	Ikawa	62	1957

Source: Electric Power Civil Engineering Association [60].

Presented by CEPCO

Figure 2.4.3 Ikawa Dam. (See colour plate section)

Aki Koichi and Comprehensive River Development [61]

Aki Koichi was born in the Niigata City in 1902. He graduated from university and entered the Ministry of the Interior in 1926. There, he worked on projects to improve the Kinu and Fuji Rivers. In 1937, he received a joint appointment to the Civil Engineering Laboratory of the Ministry of the Interior. He wrote "River Physiognomy" in 1944. In 1948, he was appointed to the post of secretary-general of the Resources Committee, Economic Stability Headquarters. For three years, beginning in 1951, he served as Vice-chairman of the Resources Survey Committee, Economic Stability Headquarters, as it tackled resource problems, particularly the use of water resources, in Japan. The Resources Committee and the Resources Survey Committee produced many fine reports that were used to prepare policies for the economic rehabilitation of Japan and for the rapid growth that followed.

Aki wrote, "During the war, I obtained an English report called 'Ten Years of TVA – Democracy Advances' written by Lillienthal, and was enthralled by it" and, "I was motivated to switch my studies from river flood prevention measures to the development of water resources through my contact with staff of the Natural Resources Department of GHQ of the occupation forces at that time." These comments reveal how, at that time when the newspapers and magazines were filled with discussions of democratic ideals, comprehensive river development modeled on the TVA that was evaluated as the embodiment of the new American democracy was being counted on to contribute to postwar rehabilitation and the advance of postwar democracy.

Aki was a living technological idealist and a free-intellect type of engineer. He was revered by many for his river ideology and for his character. He made major contributions to the postwar rehabilitation of Japan's economy by devoting himself to resources and technology with a rational spirit.

2.4.2.3   Increase in food production and development of agricultural land improvement policies and irrigation dams

Immediately after the end of World War II, Japan's population soared as demobilized servicemen and others returned from overseas, but economic production had collapsed and a food crisis ensued. In the large cities in particular, the residents were nearly starving and some foreign correspondents reported that several million Japanese might starve to death.

Under such chaotic conditions, increasing food production and expanding agricultural land were urgent challenges.

To meet these challenges, postwar land reform, based on the principle of the owner-farmer, was carried out. The Land Improvement Act came into force in 1949, followed by related laws that were the basis for large-scale irrigation and drainage projects. In this way, the emphasis of policies to increase food production shifted to the improvement of existing farmland.

The national rice paddy yield in 1955 set a new record of 12 million tons, showing that self-sufficiency in rice had nearly been achieved and that Japan had finally overcome its postwar food crisis.

The rise in paddy rice production owed a great deal to progress in rice cultivation technology and the construction of agricultural dams during this period. During the decade from 1946 to 1955, as shown in Table 2.4.3, 115 irrigation dams were constructed, making a big contribution to ensuring irrigation water.

During the same period, massive irrigation projects, agricultural land development projects and land-reclamation projects, which were typical postwar land improvement projects, commenced.

Through these projects, approximately 770 irrigation and multi-purpose dams were constructed during the 60-year period from the end of the war to 2004, thus developing 1,370 m³/s of new irrigation water supplies.

Table 2.4.3 Change in development of irrigation water and the number of dams constructed.

Period	Developed irrigation water (m³/s)			Number of dams constructed		
	Irrigation dam	Multi-purpose dam	Total	Irrigation dam	Multi-purpose dam	Total
1946–1955	48.02	7.46	55.48	115	12	127
1956–1965	86.33	260.94	347.27	100	43	143
1966–1975	77.72	289.84	367.56	105	60	165
1976–1985	74.09	135.91	210.00	71	44	115
1986–1995	107.60	128.71	236.31	79	46	125
1996–2004	74.44	81.43	155.87	59	37	96
Total	468.20	904.29	1,372.49	529	242	771

Source: Japan Dam Foundation [62].

2.4.3 Examples of roles of dams in Comprehensive National Land Development (Part 1) – The five large dams for flood control on the Kitakami River – [63]

2.4.3.1 The Kitakami River five-large-dams project

The Kitakami River is the largest river in the Tohoku Region of Japan, with a trunk channel flowing for 249 km from north to south as it drains a basin of 10,150 km². Along the Kitakami River, numerous floods have caused severe damage throughout history. Among these disasters, flood damage caused by Typhoon Kathleen in 1947 and Typhoon Ione in 1948 had a severe impact on society and on the economy in the Tohoku Region at that time.

Therefore, in 1951, the Kitakami River Basin was designated a specified region under the National Comprehensive Land Development Act, and a Comprehensive Development Plan for the Specified Kitakami River Area was prepared to protect the Kitakami River basin from floods and to develop resources that were in short supply in the region. This plan is called the KVA, after the TVA (Tennessee Valley Authority) that it was modeled on (Figure 2.4.4).

The KVA established three basic conditions: to implement flood prevention measures on the main course and tributaries of the Kitakami River; to develop the integrated use of various resources; and to stimulate the mining and manufacturing industry, centered on regional resources as its priority targets. The core features of this

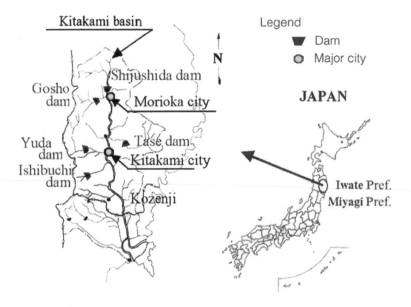

Figure 2.4.4 Kitakami River Basin and the five large dams.

Source: MLIT.

Table 2.4.4 Outline of the five large dams on the Kitakami River System [63].

Name of Dam	Ishibuchi	Tase	Yuda	Shijushida	Gosho
Name of River	Isawa	Sarugaishi	Waga	Kitakami	Shizukuishi
Type	ER	PG	PG	PG/TE	PG/ER
Catchment Area (km²)	154	740	583	1,196	635
Dam Height (m)	53	81.5	89.5	50	52.5
Total Storage Capacity (10³ m³)	16,150	146,500	114,160	47,100	65,000
Flood Control Capacity (10³ m³)	5,600	84,500	77,810	33,900	40,000
Design Flood Discharge (m³/s)	1,200	2,700	2,200	1,350	2,450
Planned Discharge (m³/s)	300	2,200	1,800	650	1,250
Power Generation (kW)	14,600	27,000	(1) 37,600 (2) 15,500	15,100	13,000
Irrigation (m³/s)	16.0	8.9	8.0	–	17.4
Municipal Water (m³/day)	–	–	–	–	64,800
Commencement/Completion	1946/1953	1941/1954	1953/1964	1962/1968	1967/1981

Source: MLIT.

Presented by MLIT

Figure 2.4.5 Ishibuchi Dam. (See colour plate section)

project were flood control and water usage projects, with five large dams as the major facilities. The project attracted considerable interest, because it not only controlled floods to protect the peoples' living environment in the drainage basin, but also ensured irrigation water to increase the production of food, which was in extremely short supply, and permitted a supply of electric energy to stimulate industry, thereby making a major contribution to the development of the Kitakami River basin.

The five large dams are, as shown in Table 2.4.4, the Ishibuchi Dam (CFRD, 53.0 m, #16, Figure 2.4.5) that was completed in 1953, followed in succession by the Tase Dam, the Yuda Dam (VA, 89.5 m, #14), the Shijushida Dam (PG/ER, 50.0 m, #11), and finally the Gosho Dam (PG/ER, 52.5 m, #12) that was completed in 1981. The dams are all managed by the MLIT.

2.4.3.2 Effectiveness of flood control

The five dams completed by the KVA have provided a combined flood control capacity of approximately 240 million m³.

The MLIT has performed the following trial calculation of the effect of the flood damage mitigation[22] of this flood control capacity, premised on the current distribution of assets and the construction of levees.

It compared the cost of damage with and without the five large dams, hypothesizing the case of the flood triggered by Typhoon Kathleen (1947) that caused postwar Japan's most severe flood damage. In the results, it was calculated that constructing the five large dams would have a damage reduction effect within Iwate Prefecture represented by about 2,900 ha of submerged land, approximately 4,800 submerged houses, and in monetary terms, damage of about 500 billion yen. A similar study, based on a simulation of recent floods, estimated that the flood damage reduction effectiveness of the five large dams would have been more than 51.3 billion yen in the 1981 flood case, over 100 million yen in the 2000 flood case, and more than 21.7 billion yen in the 2002 flood case.

2.4.3.3 Effectiveness of supplying irrigation water etc.

Four of the five dams have provided irrigation, making a big contribution to the development of agriculture in each region. The case of the Ishibuchi Dam is discussed below.

The Isawa fan that is supplied with irrigation water by the Ishibuchi Dam is an inclined fan consisting of six riverbank terraces formed a long time ago.

Therefore, the major problem in developing its agricultural land was discovering a way of raising water stably from the Isawa River to high terrace surfaces. The completion of the Ishibuchi Dam just upstream from this fan, permitted a stable supply of water to a higher elevation, and along with the construction of irrigation channels and diversion works, expanded the agricultural land area. To take the rice paddy area as an example, in 1952 it was approximately 7,800 ha, but by 1970 it had grown to about 11,700 ha (Figure 2.4.6) and the value of agricultural production climbed from about 13.7 billion yen in 1952 to about 24.7 billion yen in 1970.

The stable supply of irrigation water by the Ishibuchi Dam advanced land improvement projects and rationalized and mechanized agriculture, and along with this, it contributed to the increase in the value of agricultural production during the rapid economic growth period, with huge ripple effects on the region's economy.

---

22 Effect of the flood damage mitigation: the difference between the actual inundation damage by a flood that has actually occurred, and the hypothetical damage which would have been caused by the same flood if the completed dam had not been constructed.

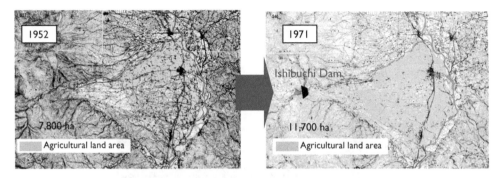

Figure 2.4.6 Change of the agricultural land area on the Isawa Fan [63]. (See colour plate section)
Source: MLIT.

2.4.3.4 The effectiveness of hydropower production

At the Ishibuchi Dam, the Isawa No. 1 Power Plant (J-POWER, 14.6 MW) was constructed, while at the Tase Dam, the Towa Power Plant (J-POWER, 27 MW) was built. The two dams, which were large enough to supply about 40% of the 102 MW of electricity produced in Iwate Prefecture in 1954, the year they began operating, played major roles in supplying electric power.

2.4.3.5 The effectiveness of local revitalization

The reservoirs created by the five dams and dam projects have played major roles as tourist resources in the region. Among these, the Gosho Dam (Figure 2.4.7), which has made a remarkable contribution to revitalizing the region around it, near the prefectural capital of Morioka City, is described in detail.

In the region surrounding the Gosho Dam, the Gosho Lake Wide Area Park Project (1980) was carried out by Iwate Prefecture to create an environmental conservation recreation base. It is surrounded by a swimming center (1985), a boating lake (1990), folklore archives (1994) and other attractions. The MLIT that manages the dams, has carried out environmental and recreational enhancement work at the reservoir (Gosho Dam Lake Park Project) at two locations, in order to use the dam as a public park.

The MLIT has reported the effectiveness of the Gosho Dam in revitalizing the region, stating that, "The number of visitors to the Tsunagi Hot Spring on the banks of the dam reservoir, which had been a sightseeing destination for many years, increased as the surrounding facilities were improved following the completion of the dam. Eventually in 1986, it reached 520,000 people, that was almost double the number before the dam was completed, and reached a record level of 840,000 people in 1991. The same trend has been seen at Koiwai Farm, another nearby tourist zone. An estimation of change over time in the amount of tourist income (amount spent by tourists) based on the number of tourists, reveals that it grew rapidly after the completion of the Gosho Dam in 1981,

Presented by MLIT

Figure 2.4.7 Gosho Dam. (See colour plate section)

rising between three and four times between 1980 (before the completion) and 1994."

With the completion of the dam, the surrounding environment has been improved and tourism is one example of how the dam has contributed to the regional economy.

2.4.4 Examples of the roles of dams in Comprehensive National Land Development (Part 2) – The Tenryu River Sakuma Dam, etc. that became the key to promoting electric power development –

2.4.4.1 Role of the Sakuma Dam in the development of the Tenryu River [64, 65, 66, 67]

The Tenryu River carries a large volume of water as a result of the heavy snow and rain that fall in its mountainous middle reaches as it flows through the Chubu Region. As a result, the region had sought development for many years dating back to the Taisho Period. Following a severe drought in 1951, electric power had to be developed very quickly, so J-POWER, which was founded in 1952, decided to develop electric power at Sakuma.

The Sakuma Power Plant was designed to handle peak loads and J-POWER also developed electric power resources at the reregulating reservoir (refer to 3.3.1.2), further downstream at Akiha.

The Sakuma Dam (PG, 155.5 m, #82, Figure 2.4.8) is a concrete gravity dam with a height of 155.5 m and reservoir capacity of about 330 million m³. The maximum output of the Sakuma Power Plant is 350,000 kW, equivalent to 2.3% of the total electric power output in Japan at that time. Its annual electric power production

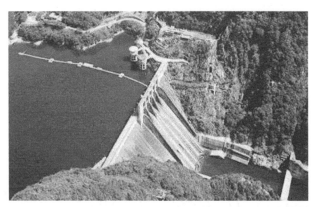

Presented by J-POWER

Figure 2.4.8 Sakuma Dam. (See colour plate section)

of about 1.5 billion kWh has been the largest in Japan from the time it was completed until now, and a pioneer in large-capacity reservoir-type hydropower plants. The power it produces has been shared by Chubu Electric Power Co. (CEPCO) and TEPCO, thus making a major contribution to stabilizing supply and demand in the Tokyo-Yokohama Area and Nagoya Area and to the economical operation of advanced thermal power plants.

The Sakuma Dam was constructed in just three years, a very short time considering the technological capabilities of the time, in order to end the shortage of electric power as quickly as possible. J-POWER replaced conventional work methods with the newest mechanized work methods, by using a tunnel excavation machine called "jumbo" to excavate the temporary diversion tunnel for example, for both the dam and power plant work, introduced large civil engineering machinery from the U.S.A. and borrowed funds from private U.S. banks to start the work as a joint venture with a U.S. company. In these ways, it completed this record-breaking work and began operating in April 1956, only three years after work started in April 1953.

As a result of this experience, Japan's public works contractors fundamentally transformed their operations, thereby modernizing their corporate organizations and labor management systems. Public works construction machinery manufacturers also introduced foreign technologies, and beginning with the construction of the Sakuma Dam, domestic manufacturing of large construction machines began within a few years. New hydroelectric technology or work methods introduced to execute this work stimulated the rapid advance of the construction and construction machinery industries. This project had a progressive impact on the labor market and local community by constructing roads, providing public facilities, and strengthening the finances of local governments.

To support the postwar policy of increasing food production in the Tenryu and Higashi-Mikawa Regions, comprehensive development projects, including the comprehensive development of the 7,000 ha-wide Mikatahara Plateau, were carried out over a wide area around the downstream section of the Tenryu River, where ensuring

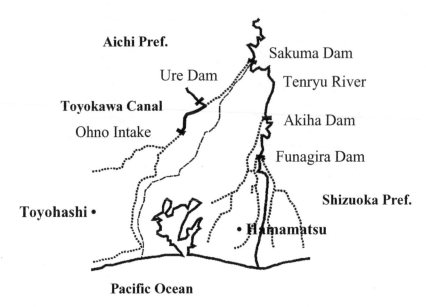

Pacific Ocean

Figure 2.4.9 Water supply systems in the Tenryu and Higashi-Mikawa Regions.

Source: M. Kubota [68].

water for irrigation, waterworks and industry were major challenges. These plans were integrated with electric power development projects focused on the Sakuma Dam, and in 1951, the Tenryu and Higashi-Mikawa Regions were designated as specified regions under the National Comprehensive Land Development Act. As shown in Figure 2.4.9, water supplied to the above development projects was taken in and supplied by the Akiha Dam (PG, 89.0 m, #83) and the Sakuma Dam.

2.4.4.2 Cases of regional development of large-scale reservoir-type hydropower dams

In addition to development under comprehensive development projects based on the National Comprehensive Land Development Act, in many cases regional development accompanied the construction of large-capacity reservoir type hydropower dams. Examples of this include cases where, in the light of the geological fact that dams are constructed in mountains, with the participation of concerned local governments, new towns are created premised on providing revitalizing measures for resettlement and compensation for public facilities. Others are cases where newly-constructed roads that are built to transport construction materials needed for dam and power plant work are available to the region after the work has been completed and utilized for regional development and tourist resources.

When the Ikawa Dam (HG, 103.6 m, #77) on the upstream part of the Oi River was constructed, the planned site of the power plant was an isolated mountainous region that was extremely inaccessible. So CEPCO constructed a specialized railway to transport construction materials. After the work was completed, the specialized railway was provided for use as transportation for people living along the river and as a replacement for running logs down the river, which became impossible because of the construction of the dam. It has been operated under a management contract as the Oigawa Line on the Oigawa Railway since 1959 [69]. The Nagashima Dam (PG, 109.0 m, #78) project, implemented thereafter, replaced a partial section of railway with Japan's only Abt system track that has been actively used as a tourist resource.

2.4.5   Examples of roles of dams in Comprehensive National Land Development (Part 3) – The Kiso River Makio Dam for irrigation –

2.4.5.1   Outline of the Aichi Canal Project [70, 71] (Figure 2.4.10)

The Chita Peninsula in southern Chubu Region has been a farming region since ancient times, but as a hilly zone it had almost no rivers, so its paddy fields had to rely on irrigation ponds and its dry fields on rainwater. Obtaining water from the Kiso River with its heavy flow rate as a solution was first suggested in the late Meiji Period, but transporting water over more than 100 km was impossible at that time. Under these circumstances, severe droughts that struck in the summers of 1944 and 1947 increased demand for a way to transport water from the Kiso River to the Chita Peninsula [72].

This was the situation when, in 1951, the Kiso River basin was designated as a specified region under the National Comprehensive Land Development Act and the Aichi Canal Project began.

The Aichi Canal Project was the first large-scale comprehensive development project after World War II that included both electric power production and municipal water supply. This project was undertaken with the primary objective being to construct the Makio Dam (ER, 105.0 m, #66) on the upstream Kiso River, take in the water discharged by this dam along with part of the natural flow of the Kiso River and carry this in a channel to supply irrigation water to approximately 30,000 ha of farmland, including the whole of the Chita Peninsula. In addition, about 45 million m³ of water for waterworks and for industrial use is supplied to Nagoya City and other regions where it is needed. At the same time, the water stored by the dam is used to increase electric power production.

This project was, because of financial circumstances at that time, undertaken with loans from the World Bank and was executed in the short period of five years from September 1957 to September 1961 by the Aichi Irrigation Public Corporation (that later merged with the JWA), which was established in October 1955 [73].

2.4.5.2   Changes in the plan of the Aichi Canal Project (Table 2.4.5)

The Aichi Canal Project initially supplied 28.60 m³/s of irrigation water, 1.01 m³/s for waterworks, and 0.69 m³/s of industrial water with the Makio Dam (Figure 2.4.11) as the principal water resource. At the same time, it supplied a total of 130 million kWh of electricity per year from the Mio Power Plant (34 MW) and from another power

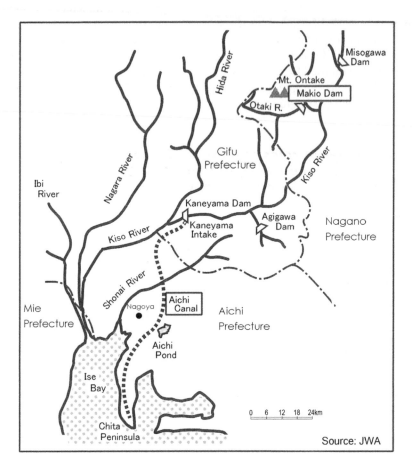

Figure 2.4.10 Outline of the Aichi Canal Project.

Source: JWA.

plant downstream. The Makio Dam's effective reservoir capacity of 68 million m^3 was about 80% of the total water resources capacity of 84.35 million m^3 of the entire Aichi Canal Project.

The Aichi Canal Project continues to play its role as a major artery for water that supports the daily life and production of the region, and has played a major part in the sweeping progress achieved in the Chubu economic region. However, as the years have passed, a decline in the beneficiary agricultural area has occurred because of the urbanization of farming villages near cities, the expansion of the region supplied by waterworks and increased water demand resulting from the rising population. The growth of industrial water demand in the portside industrial zone in southern Nagoya and other factors have created new water demand and caused changes in the forms of water demand. Additionally, the channel facilities have deteriorated, the surrounding land was converted to housing use and problems with ensuring safety of the system have appeared.

Table 2.4.5 Change in the plan of the Aichi Canal Project.

Changes of Plans Items	Original Plan (Oct. 1957)	1st Revised Plan (Oct. 1961)	2nd Revised Plan (Sep. 1964)	3rd Revised Plan (Mar. 1968)
Irrigation Beneficiary Area (ha)	33,100	30,700	23,500	15,000
Necessary Water Volume (10^6 m³/year)				
Irrigation Water	110	150	111	72
Municipal Water	18	23	23	63
Industrial Water	27	22	116	202
Capacity of Dam Reservoir (10^3 m³)				
Makio	68,000	68,000	68,000	68,000
Matsuno-ike	2,350	2,350	2,350	2,350
Togo-ike	9,000	9,000	9,000	9,000
Souri-ike	–	–	5,000	5,000
Annual Power Generation (MWh/year)	105,535	130,000	127,178	124,249

Source: MAFF [74].

To resolve these problems, the Aichi Canal Second Phase Project, which started in 1983, radically reconstructed the deteriorated Aichi Canal facilities to restore their functionality and reinforce their safety. At the same time, it ensured new water resources to meet the rising demand for water for urban use, stabilized the water supply and raised the level of water use to expand channel functions and modernize management facilities. During the Second Phase Project, the Agigawa Dam (ER, 101.5 m, #69) and the Misogawa Dam (ER, 140.0 m, #64) that were constructed by other projects increased water resources so that water rights were 21.51 m³/s of irrigation water, 6.47 m³/s for waterworks, and 9.42 m³/s of industrial water. A large mountainside failure on Mt. Ontakesan, upstream from the Makio Dam, which was caused by the Western Nagano Prefecture Earthquake in 1984, triggered the continuing inflow of large quantities of sediment. Therefore, in 1996, Measures to mitigate the sedimentation of the Makio Dam were added to the Aichi Canal Second Phase Project in order to restore the functionality of the reservoir and prevent disasters in the surrounding region.

2.4.5.3 Roles played by the Makio Dam

The Makio Dam has ensured water resources that have permitted irrigation water to be transported through the Aichi Irrigation System to the tip of the Chita Peninsula where there are no usable rivers, thus developing diverse agriculture in this suburban region and increasing the value of agricultural production. Since the Makio Dam was completed, it has supplied industrial water, which is

Presented by JWA

Figure 2.4.11 Makio Dam. (See colour plate section)

indispensable as a resource to expand industrial complexes along the coast of the Chubu Region and a driving force behind the growth of industry in Japan. It has also provided a stable supply of water valuable for daily life to support the soaring population.

The effectiveness of the Aichi Canal Project that was mainly intended to ensure water resources at the Makio Dam, is shown to be as follows, based on a comparison of the situation before (1963) and after (2000) the commencement of operations [75].

2.4.5.3.1   Development of diverse agriculture

Ensuring irrigation water has not only ended drought damage in paddy fields, it has also created a variety of high-quality and highly-productive agricultural activities by irrigating dry field regions. This has increased the gross value of agricultural production by about three times to approximately 70 billion yen, with the production of vegetables, fruit, and flowering plants increasing remarkably.

2.4.5.3.2   Advancing urbanization

The supplied population of the Aichi Canal Project has increased about six-fold, to approximately 1.2 million people.

2.4.5.3.3   Advancing industrialization

The value of the shipments of manufactured products in the region supplied by the Aichi Canal Project has increased about 13 times, to approximately 4 trillion yen.

2.5   DAMS THAT HAVE SUPPORTED THE CONCENTRATION OF POPULATION AND INDUSTRY IN LARGE CITIES AND THE ADVANCE OF NATIONAL URBANIZATION

From the time of postwar rehabilitation to the present, dams have met the rising water demand of cities, where people and industry were concentrating, thus responding to soaring electric power demand during the rapid economic growth period, and ensuring safety from flooding in regions that were urbanizing rapidly.

2.5.1   Rapid economic growth and Comprehensive National Development Plans

The Japanese economy, which had been dealt a severe blow by World War II, had recovered to its prewar level by the first half of the 1950s. During the late 1950s, the central government restored prosperity by enacting economic plans, including the Five-year Program for Economic Independence (1956) and the New Long-term Economic Plan (1958). Then, under the National Income Doubling Plan that the central government enacted in 1960, water resources development was incorporated in the national economic plan for the first time. As basic policies, it advocated that the water supply required to meet rising water demand be achieved by using reservoirs to stabilize the flow and increase the water usage, that a system for advanced development was necessary and that flood control and water use should be organically linked.

During the postwar rehabilitation period and the subsequent rapid growth period, population and industry were continually concentrated along the Pacific Belt Zone from the Metropolitan Tokyo Area to the northern part of Kyushu Island, as shown in Figure 2.5.1, thereby increasing the severity of social problems posed by the gap between this Pacific Ocean coastline and other parts of Japan. It was feared that the implementation of the National Income Doubling Plan would widen this gap.

Under these circumstances, the First Comprehensive National Development Plan that was enacted by the central government in 1962 aimed to establish development centers throughout Japan and to stimulate regions with the ripple effects emanating from these centers in order to achieve balanced development by narrowing the gaps between regions. However, during the later period of rapid growth of the Japanese economy, the problem of overcrowding in urban areas accompanied by depopulation in rural areas become increasingly severe, prompting the enactment of the New Comprehensive National Development Plan (New Development Plan) in 1969. Under the New Development Plan, a principal axis extending for 2,000 km from the north to the south of the Japanese archipelago was formed through the large-scale project development approach in order to respond to revolutionary technological progress and national urbanization by energizing the long-term, sustained and rapid development of national land. The large-scale development project approach meant combining the provision of national networks of expressways, the Shinkansen (bullet train), and communication networks with the development of large scale industrial infrastructure at key points.

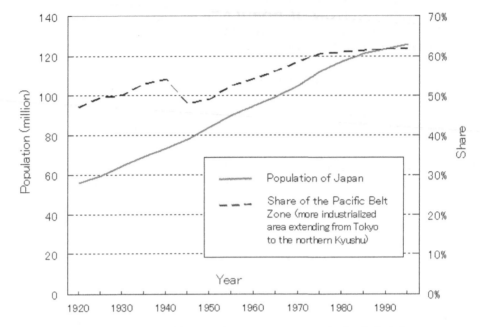

Figure 2.5.1 Change in population of Japan and its concentration in the Pacific Belt Zone.
Source: Ministry of Internal Affairs and Communications.

In these plans, systematically developing new water resources over wide areas to en-sure stable supplies of water was viewed as an extremely important challenge, reflect-ing the government's awareness that in the future water demand would soar as a result of the growing use of industrial water spurred by the spectacular growth of industrial production, the expanded use of irrigation water in response to the modernization of farm management, and rising demand for supplies by waterworks to respond to the expanding supplied population and improvement in people's lifestyles.

As the public increasingly demanded that flood control measures be taken on riv-ers and that reservoirs be used to regulate flood discharge, they advocated that future reservoir development be based on a multi-purpose approach that would integrate flood control and water usage.

2.5.2 Supplying water to large cities

2.5.2.1 *Rising water demand in large metropolitan areas*

During the postwar period of rapid growth, the water demand for waterworks climbed steadily to keep pace with the growing population and the increasing coverage of water-works systems. Japan's population was 83.2 million in 1950, and by 1960, it had risen to 93.42 million: an increase of about 10 million people in only 10 years. The water-works coverage rate grew from only 32.3% in 1955 to 53.4% by 1960, indicating that more than half of the population then used water supplied by waterworks. This means that the population being supplied with water increased by 21 million people during a period of only five years beginning in 1955. Such an increase in the population and coverage pushed up the waterworks' demand from 2.244 billion m^3 in 1951 to 2.907 billion m^3 in 1955, and then to 4.166 billion m^3 in 1960 [76] (Figure 2.5.2).

Furthermore, in the large metropolitan areas, population concentration, a rising municipal water coverage, an increase in the quantity of water consumed per person per day caused by lifestyle modernization and water usage for urban activities in business and commercial districts (restaurants/bars, department stores, hotels and other commercial water usage, water use in offices, public use of water in park fountains, public toilets, etc., and other factors), triggered a rapid rise in the waterworks' demand for water.

To satisfy this rise in water demand, many factories, buildings and waterworks organizations sought water sources in groundwater, which is good quality water obtainable at relatively low cost. One result of this was severe ground subsidence across Japan. (Figure 2.5.3) For example, on low alluvial land in Tokyo, cumulative ground settlement from 1918 to 1960 exceeded 3 meters, not only damaging buildings and roads, but also exacerbating the damage caused by urban floods. As a result, strict groundwater restrictions[23] were enforced, so that as the quality of groundwater continued to decline, the need to develop river water to meet rising demand increased at once [77].

2.5.2.2 Establishment of the water resources development promotion system and the growing water supply

Water resources development to meet such a rapid rise in water demand had to be achieved comprehensively and efficiently from a long-term perspective because of topographical and geological limits on sites suitable for dam reservoirs, and because such projects are very time-consuming.

Therefore, in 1961, the Water Resources Development Promotion Law was enacted. Under this law, when it becomes necessary to implement emergency action for wide-area water supply measures to respond to the development and growth of industry and the rising urban population, the river systems in this region are designated as Water Resources Development River Systems and a Water Resources Development Basic Plan (WRDBP) is prepared. Such plans stipulate the prospective water demand in each river system and items concerning the construction of facilities needed for these river systems to supply water, and prescribe the projects undertaken to develop and utilize the region's water resources. The central government (National Land Agency of the Prime Minister's Office) designates river systems and prepares each WRDBP. It also established the Water Resources Development Public Corporation in 1962, as a project implementation body. Financially, this corporation could apply treasury investment and loan funds to projects, in addition to its normal budget.

---

23 Groundwater restrictions: In cities that are dependent on groundwater resources, ground settlement has occurred as a result of excessive pumping of groundwater to meet rising water demand. Measures have been taken to deal with this problem by switching to the use of surface flow water as a water resource and restricting the pumping of groundwater, resulting in the stabilization of ground settlement in recent years.

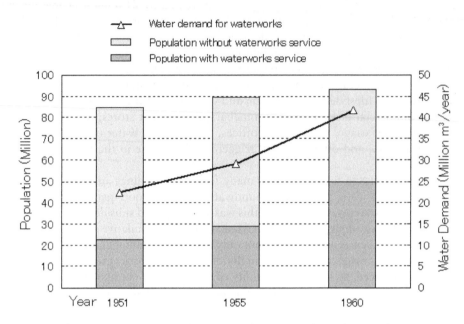

Figure 2.5.2 Changes in water demand for waterworks and served population in the post-war era.
Source: S. Yamamoto [76].

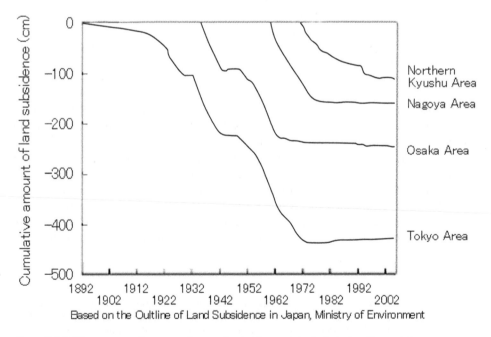

Figure 2.5.3 Change over the years of ground subsidence at typical observation points.
Source: MLIT [78].

In April 1962, the Tone and Yodo River Systems were designated as Water Resources Development River Systems, followed by the Chikugo River System in 1964, the Kiso River System in 1965 and the Yoshino River System in 1966, and WRDBPs were drawn up for each river system.

In 1974, the Ara River System was designated and the basic plans for the Tone and Ara River Systems were integrated. Then in 1990, the Toyo River System was designated as the seventh Water Resources Development River System. The WRDBPs for each river system have been reviewed and revised regularly, based on changes in economic and social conditions discovered during surveys carried out to predict supply and demand.

Regions that are supplied with water under a WRDBP account for about 51% of Japan's total population and are the origin of about 47%, by value, of all industrial shipping in Japan.

As a result of water resources development done throughout Japan by the central government, local governments and the Water Resources Development Corporation, the percentage of water taken from dams through waterworks projects rose by more than three times, from 11.6% in 1965 to 38.5% in 1999, as shown in Table 2.2.3.

The Water Resources Development Corporation was reorganized under a new name, the Japan Water Agency (Incorporated Administrative Agency, JWA), in 2002.

The former Water Resources Development Corporation and JWA developed a total capacity of about 335 m³/s up to 2004, based on the WRDBPs for the various river systems [79].

2.5.2.3 Examples of multi-purpose dams supporting the concentration of population and industry in large metropolitan regions

2.5.2.3.1 Progress of dam projects on the Tone and Ara River Systems

The progress of dam projects on the Tone and Ara River Systems, which are major sources of water for Metropolitan Tokyo, is described below.

Water usage on both river systems has been primarily agricultural since ancient times, but from the Meiji Period until the 1930s, their water was increasingly used to supply municipal water and to generate electricity. As Japan recovered after the end of World War II, full-scale water development got underway. In 1950, work began on the Ikari Dam (completed in 1956; numbers in brackets in the remainder of this paragraph show the year of completion), followed by the construction of the Fujiwara Dam (1958), the Aimata Dam (PG, 67.0 m, #38, 1959), the Sonohara Dam (PG, 76.5 m, #36, 1965), the Kawamata Dam (VA, 117.0 m, #27, 1965), the Futase Dam (VA, 95.0 m, #42, 1961), and other multi-purpose dams that supply existing irrigation (refer to 2.4.2.1), control floods and generate electricity, were constructed as national government projects.

In Tokyo, with the arrival of the rapid economic growth period, the population exceeded 9 million in about 1960 and the concentration of industry continued, resulting in soaring water demand accompanied by severe water shortages, resulting in the region being called the Tokyo Desert.

Specifically, beginning in 1961, a prolonged drought on the Tama River, which had a long history as a water source for Tokyo, forced restrictions on the supply of water from the river in successive years. In May 1962, the Temporary Tokyo Drought Countermeasure Headquarters was established. Later, temporary water supply restrictions were relaxed, but beginning at the end of 1963, water supply restrictions were resumed and during the summer of 1964, immediately before the Tokyo Olympics (1964), the quantity of water stored in the three reservoirs of the Ogochi Dam, the Murayama Dam and the Yamaguchi Dam, had fallen to 1.6% of their combined capacity, forcing the fourth water supply restrictions to be enforced (the target reduction was 50%, a restriction that included cutting off daytime water supply) [80].

In order to resolve such problems, the Tone River System was designated as a Water Resources Development River System in 1962, and in August of the same year, its WRDBP was finalized and announced. This plan focused on the early completion of the Yagisawa Dam (VA, 131.0 m, #31, Figure 2.5.4) and the Shimokubo Dam in order to obtain a new water supply of 30 m³/s that could guarantee water as quickly as possible.

The Tone River System WRDBP was revised by the addition of the Tone Canal, the Imbanuma Development (#46) and the Gunma Canal Projects in 1963. Then in 1964, the so-called First Tone River WRDBP, which enacted a water demand and supply plan with 1970 as its target year (Figure 2.5.5), was established.

The First Tone River WRDBP estimated that water demand would rise to a total of about 120 m³/s in fiscal 1970, which included 50 m³/s of water needed to expand waterworks, 30 m³/s of water needed to develop industry and 40 m³/s of water needed to develop farmland and modernize agriculture. To deal with this new water demand, it set the goal of supplying about 120 m³/s by constructing facilities to develop and utilize water resources, and facilities to utilize the existing water supply system

Presented by JWA

Figure 2.5.4 Yagisawa Dam. (See colour plate section)

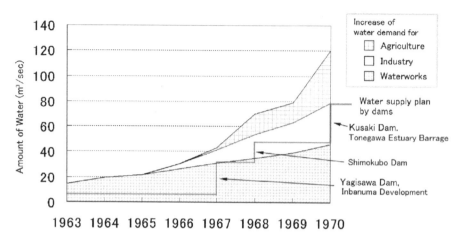

Figure 2.5.5 Water demand and supply plan in the First Tone River WRDBP.
Source: JWA.

more rationally, and at the same time, to take measures to utilize water resources rationally [81].

Supply facilities that were needed to achieve these goals were, in addition to the Yagisawa Dam, Shimokubo Dam and the Imbanuma development, which were already considered in the basic plan, the Godo Dam (later renamed the Kusaki Dam) and the Tonegawa Estuary Barrage (Weir, #45) that had already been surveyed. These were developed as water resources while the Tone Canal was positioned as a water conveyance facility.

The Tone Canal Project, which started in 1963 and was completed in 1968, was the first project undertaken independently by the JWA. In 1964 and 1965, while the project was under construction, a severe drought struck Metropolitan Tokyo, so the incomplete facilities were used to supply water to Tokyo as an emergency measure, thus avoiding a water shortage during the Tokyo Olympics.

On the Tone River System, the Naramata Dam and the Omoi River Development (Namma Dam (CFRD, 86.5 m, #29)), the Kawaji Dam (VA, 140.0 m, #24) and the Yamba Dam (PG, 131.0 m, #39) and others were added to meet future water demand. In 1974, the Ara River System that adjoins the Tone River System was designated as a Water Resources Development River System, and the Takizawa Dam (PG, 140.0 m, #41) and the Urayama Dam (PG, 156.0 m, #43) were planned (Figure 2.5.6).

Improvements under the plan continued, so that residents of Tokyo are now dependent on dams constructed under this WRDBP for 70% of their water. As a result, the group of dams on the upstream Tone River replenishes water for more than 200 days per year, so that the river will always supply the required quantity of water. For example, in the summer of 2004, intense heat and low rainfall struck Metropolitan Tokyo, but at the same time, dams effectively stored rain falling in their catchments and midstream reservoirs, downstream aqueducts and estuary weirs were

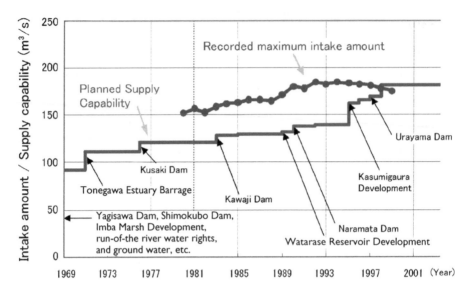

Figure 2.5.6 Water demand and supply on the Tone and Ara River Systems (municipal and industrial water supply).

Source: MLIT.

used to perform excellent water management, thereby successfully avoiding water restrictions (Figure 2.5.7).

2.5.2.3.2   The progress of dam projects on the Yodo River System fed by Lake Biwa

The use of groundwater in response to rising water demand in the Kyoto-Osaka-Kobe Metropolitan Area caused serious ground settlement phenomena on low-lying land along the downstream Yodo River, increasing the population's expectations of river water development.

The Yodo River System WRDBP was settled in 1962 and then partially revised in 1976. Under the plan, the rise in water demand in 1980 fiscal year was predicted to be about 68 m³/s (43 m³/s for waterworks, 23 m³/s for industrial use, and 2 m³/s for agricultural use), so the targeted supply to meet this need was set at 50.4 m³/s [81]. A feature of the Yodo River WRDBP is that 80% of the water is developed by the Lake Biwa Development Project (#88, 40 m³/s developed), and as facilities to supply water resources in the target year, five other dams – the Murou Dam (PG, 63.5 m, #97), the Hitokura Dam (PG, 75.0 m, #91), the Hiyoshi Dam (PG, 70.4 m, #89), the Hinachi Dam (PG, 70.5 m, #87) and the Nunome Dam (PG, 72.0 m, #96) – were planned and constructed.

The Lake Biwa Development Project prevents inundation damage along the shoreline of Lake Biwa, which is Japan's largest lake, and permits the development

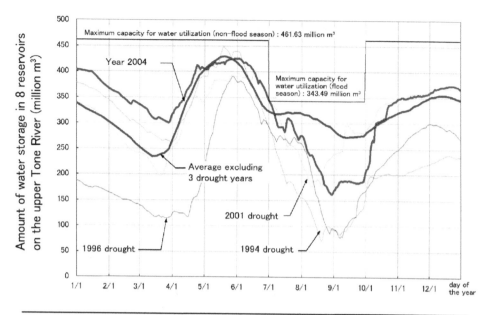

Replenishment by 8 dams on the Upper Tone River (July)

| Year | Past drought years | | | Average (1992–2003, excluding 3 drought yeas) | 2004 |
	1994	1996	2001		
Replenished amount (million m³)	146.47	62.35	117.59	17.88	109.71

Figure 2.5.7 Replenishment by dams on the Upstream Tone River in 2004. (See colour plate section)
Source: MLIT.

of water resources with its reservoir capacity. At the same time as the project was implemented, wide area regional development projects, including the construction of sewage systems, were executed in the water source region, which is the land along the shores of Lake Biwa. The projects are referred to collectively as the Lake Biwa Comprehensive Development Project. The project's framework was established by the Law for Special Measures for the Lake Biwa Comprehensive Development in 1972.

The Lake Biwa Development Project developed new water resources of 40 m³/s by constructing lakeside dikes and increasing the flow capacity of the Seta River (upper reach of the Uji River), which is the only river that discharges from Lake Biwa, and at the same time regulates the quantity discharged from the Seta River Weir (Figure 2.5.8), thus permitting the use of Lake Biwa water within a water level range of +0.3 m to −1.5 m. After two extensions to the work period, the project was completed in 1996.

Figure 2.5.8 Seta River Weir. (See colour plate section)
Source: MLIT.

2.5.2.3.3   The progress of dam projects on the Yoshino River System on Shikoku Island

Water resource development on the Yoshino River began in earnest with the enactment of the Yoshino River Comprehensive Development Project in 1966, based on
the Regional Development Promotion Law. Under the Yoshino River Comprehensive
Development Project, with the Sameura Dam as the core facility for water usage,
flood control and electric power production, the Ikeda Dam (PG, 24.0 m, #108), the
Kagawa Canal water supply system that diverts and conveys water from the Yoshino
River to the Sanuki Plain outside its drainage basin, the Shingu Dam (PG, 42.0 m,
#111) and the Kyu-Yoshinogawa Estuary Barrage (Weir, #107) and diversion facilities
that conveyed water to the Kochi Plain outside the drainage basin, were constructed.
Similarly, in 1966, the Yoshino River System was designated as a Water Resources De-
velopment River System and construction was completed by JWA in 1978. Then the
Tomisato Dam (PG, 111.0 m, #113) was completed and the seven projects produced
922 million m³ annually, which was shared among the four prefectures on Shikoku
Island [82].

2.5.2.4   The roles of dams in supporting the growth of regional cities

The provision of dams and other water resources development facilities provided a
stabilized supply of water for urban use, and nationally, in 1995, water resources
provided by dams stood at 14.1 billion m³ per year, thus providing 43% of urban use
water (Figure 2.5.9).

Urban water use ensured by dams and other water resources development facilities
supported the growth of regional cities where water demand soared as a result of population concentration and the spread of sewage systems during the rapid economic
growth period.

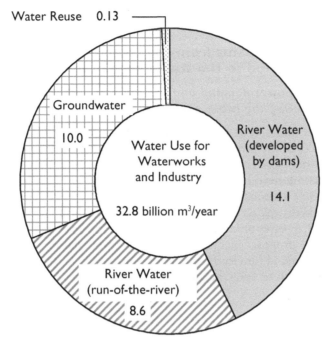

Figure 2.5.9 A breakdown of water resources supply (1995).
Source: MLIT [83].

In Morioka City in Iwate Prefecture in the Tohoku Region (the Kitakami River System) for example, water supply systems were expanded, with the Gosho Dam (MLIT, 1981) as their water source, to meet rising demand. Then the Fifth Waterworks Expansion Project (target year 1975) added a capacity of 10,000 m³/day from the intake facilities located 8 km downstream from the dam, and the Seventh Waterworks Expansion Project (target year 1985) added another 54,800 m³/day by direct intake from the dam reservoir, for a combined increased intake of 64,800 m³/day. Approximately 43% of the water supply capacity at that time (151,700 m³/day) was provided by the Gosho Dam [84].

There are many regional cities where rising urban water demand resulted in restrictions on the water supply during droughts, with a severe impact on urban life. In Fukuoka City in Kyushu for example, water supply had to be restricted for 287 days in 1979 and for 64 days in 1994. In Takamatsu City in Shikoku Island it was restricted for 58 days in 1973 and 67 days in 1994, and in Naha City in Okinawa Island, for 239 days in 1973 and for 326 days in 1981.

In response to such drought disasters and rising urban water usage, multi-purpose dams were constructed for Fukuoka City: the Egawa Dam (JWA, 1972) on the Chikugo River System, for Takamatsu City: dam groups including the Sameura Dam under the Yoshino River Comprehensive Development Project, for Naha: the Fukuji Dam (ER, 91.7 m, #135) and other dams in the Okinawa North Dam Group. These dams have greatly helped the residents of these drought-plagued regions enjoy more stable lives.

2.5.3 Hydropower dams from the rapid economic growth period to the stable growth period

2.5.3.1 Electric power demand and the roles of hydropower dams during the rapid economic growth period

The development and spread of household electrical appliances was a particularly remarkable aspect of the process of postwar economic growth in Japan, as washing machines, television sets and refrigerators spread rapidly in homes during the late 1950s. This was followed by a revolution in consumption of the so-called three Cs: cars, coolers (home air-conditioners) and color TV.

Economic growth was accompanied by a rapid growth in energy demand. Beginning about 1948, a series of large-scale oil fields were discovered in the Middle East and technological progress encouraged a switch from coal to petroleum in the industrial world. Demand for electric power also increased so that, as shown in Figure 2.5.10, from the 1950s to 1965 the quantity of electric power consumed soared almost

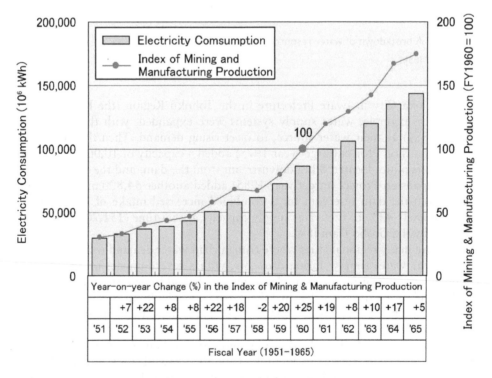

Figure 2.5.10 Change in the index of mining and manufacturing production and the quantity of electricity consumption between the end of the War and 1965.

Source: Ministry of Economy, Trade and Industry (METI) [85].

as rapidly as the annual 10% increase in the mining and manufacturing industry's production index. During this period, electric power companies were compelled to ensure energy supplies by means of large-scale electric power source development. Figure 2.5.11 shows changes in the state of hydropower output excluding pumped-storage type hydropower increased significantly, and especially during a period of more than ten years beginning in about 1955.

Table 2.5.1 shows examples of dams for large-scale reservoir type electric power plants following the Sakuma Dam, which was constructed prior to the rapid economic growth period. These were the dams that created the golden age of hydropower development.

The hydro-first/thermal-second electric power structure continued until 1962 and was followed by the advance of thermal power and nuclear power, but even after 1960, reservoir and regulating pond type hydropower plants continued to be developed as valuable peak-supply power. The concept of river hydropower development is to construct groups of hydropower plants appropriately from upstream to downstream to efficiently produce hydropower from the overall river perspective, and is, accordingly, called Consistent Hydropower Development in a River System. The oil shock of 1973 was followed by large-scale pumped-storage electric power production as part of valuable clean energy and as power to respond to peak electric power production.

2.5.3.2   The redevelopment of hydropower by Consistent Hydropower Development in a River System

2.5.3.2.1   An outline of Consistent Hydropower Development in a River System with the aim of raising hydropower production

On major river systems in Japan, hydropower development was started in the 1920s by constructing dams in the central and downstream reaches of rivers, where it was easy to construct electric power plants. Thereafter, large-scale hydropower development shifted upstream. From about 1960, the construction of hydropower plants resumed from the upstream reaches to the central and downstream reaches of each river system, and electric power plants were developed or redeveloped in the central and downstream reaches of rivers to efficiently utilize the head drop and water quantity. Good examples are the Kiso River, Hida River, Oi River, Agano River, Sho River and Kurobe River.

Below, the Kurobe River, which is the location of the Kurobe Dam (VA, 186.0 m, #58, Figure 2.5.13), the highest dam in Japan, is introduced as an example of Consistent Hydropower Development in a River System.

2.5.3.2.2   Consistent Hydropower Development in a River System to increase electric power production on the Kurobe River

As explained in 2.3.5, until the 1940s, electric power plants were constructed on the Kurobe River in a series of steps, thus taking advantage of the head drop of river water from the old Yanagawara Power Plant to the Kurobegawa No. 3 Electric Power Plant (Table 2.5.2, Figure 2.5.14).

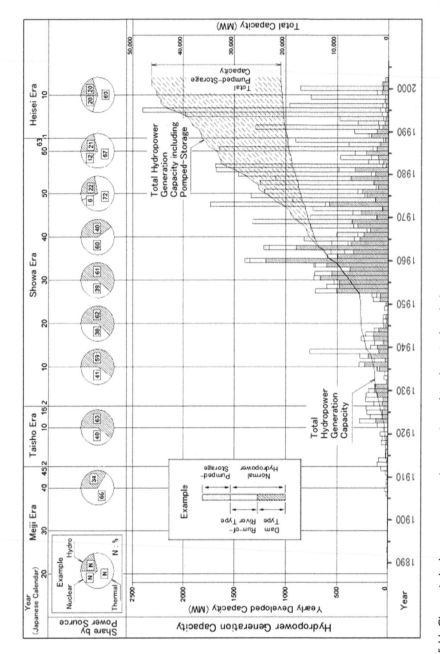

Figure 2.5.11 Change in hydropower generation capacity and its share in the whole generation capacity.

Source: Electric Power Civil Engineering Association [80].

Table 2.5.1 Hydropower dams completed around 1960.

Name of Dam	Owner	River System	Dam height type	Name of Power Plant	Output Power	Start of Operation
Tagokura	J-POWER	Agano	145 m PG	Tagokura	380 MW	1959
Okutadami (Figure 2.5.12)	J-POWER	Agano	157 m PG	Okutadami	360 MW	1969
Miboro	J-POWER	Sho	131 m ER	Miboro	215 MW	1961
Kurobe	KEPCO	Kurobe	186 m VA	Kurobegawa No. 4	335 MW	1960

Source: Electric Power Civil Engineering Association [87].

Presented by J-POWER

Figure 2.5.12 Okutadami Dam. (See colour plate section)

After World War II, an age when electric power production had shifted to advanced thermal power while hydropower played a role as a large reservoir-type peak load supply, the KEPCO responded by preparing a plan to construct a dam to form a large reservoir at the furthest upstream part of the Kurobe River, which was to play a pivotal role in the Consistent Hydropower Development in a River System . The Kurobe Dam, which attracted attention as one of the century's giant projects, was completed in 1963 and was the product of the finest civil engineering technology in Japan at that time. The Kurobegawa No. 4 Hydropower Station that takes water from the reservoir at the Kurobe Dam, began operating in 1961, prior to the completion of the Kurobe Dam.

Presented by KEPCO

Figure 2.5.13 Kurobe Dam. (See colour plate section)

Table 2.5.2 Consistent hydropower development on the Kurobe River.

Type of development	Completion	Power Plant (Dam)
Development in the lower reach	1927–1947	Yanagawara P.S. Aimoto P.S. Kurobegawa No. 2 P.S. (Koyadaira Dam) Kurobegawa No. 3 P.S. (Sennindani Dam) Kuronagi No. 2 P.S.
Large-scale reservoir development in the upper reach	1961	Kurobegawa No. 4 P.S. (Kurobe Dam)
Redevelopment in the lower reach	1963–1985	Shin-Kurobegawa No. 2 P.S. (Koyadaira Dam) Shin-Kurobegawa No. 3 P.S. (Sennindani Dam) Otozawa P.S. (Dashidaira Dam)
	1993–2000	Shin-Yanagawara P.S. (Dashidaira Dam) Unazuki P.S. (Unazuki Dam)

Source: Electric Power Civil Engineering Association [88].

The novel, The Kurobe Sun (1966), is a literary work that was later made into a film. The book realistically portrays the engineers who challenged the power of nature and their own humanity through the construction of the Omachi Tunnel, which was a transport road used to construct the Kurobe Dam. The people and their struggle with raw nature portrayed in this movie, provided courage and excitement, attracting large audiences and motivating many young people to study civil engineering.

The completion of the Kurobe Dam with its total reservoir capacity of approximately 200 million m³ improved the downstream flow regime remarkably. To use its capacity effectively, the New Kurobegawa No. 3 Power Plant (1963) and the New Kurobegawa No. 2 Power Plant (1966), located downstream, were constructed in succession. Then the Otozawa Power Plant (1985) was constructed, completing the entire Consistent Hydropower Development on the River (Figure 2.5.14). In this way, the Kurobe River became a power-source river with a series of peak power plants that took full advantage of the head drop of more than 1,300 m from the Kurobe Dam reservoir water level (elevation 1,448 m) to the Otozawa Power Plant (elevation 131.1 m).

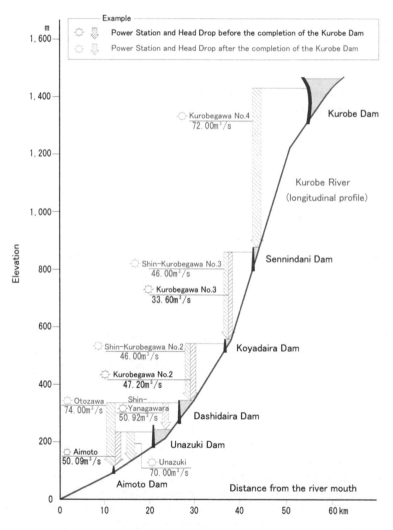

Figure 2.5.14 Schematic view of hydropower generation on the Kurobe River. (See colour plate section)

Source: KEPCO.

The MLIT reacted to severe flooding in 1969 by planning a specific multi-purpose dam for flood control, waterworks supply and electric power production. With the construction of this dam, named the Unazuki Dam (PG, 97.0 m, #54), the Yanaga-wara Power Plant was reconstructed in 1993 as the New Yanagawara Power Plant. Thereafter, the Unazuki Power Plant was completed, raising the total output of ten power resources developed by KEPCO on the Kurobe River System to 890,700 kW (Table 2.5.2).

At the Kurobe Dam, water produced by melting snow is stored from the spring to the summer and is used in the winter, from December to March, by power plants downstream from the Kurobegawa No. 4 Power Plant, contributing to a stable supply of electric power [89].

The storage capacity of the Kurobe Dam controls flooding along the downstream section of the river by, for example, mitigating flood runoff thanks to the huge capacity of its reservoir, and at the same time, it creates new tourist attractions, contributing to regional economic development and to tourism development.

2.5.3.3   Hydropower development centered on pumped-storage type hydropower

2.5.3.3.1   Background to the development of pumped-storage hydropower [90]

In the late 1950s, high-capacity, advanced thermal power plants took over the base load supply of electric power, with peak adjustment handled by large-scale reservoir-type hydropower plants.

During the 1960s, rapid urbanization and a rise in the people's standard of living driven by rapid economic growth resulted in a remarkable increase in office and home electricity demand for air-conditioners. This trend shifted the annual maximum power demand, which had formerly been on winter evenings, to the daytime during summer. The summer peak exceeded the winter peak in 1968.

This summer peak created a new demand pattern, marked by a sharpened peak during the day, a pattern that was beyond the adjustment capacity of reservoir-type power plants, thereby creating a need for pumped-storage power plants that are better suited to adjusting the gap in daytime and nighttime electric power demand.

In 1960, the Resources Council of the Science and Technology Agency of the Prime Minister's Office issued a policy statement calling for the diversification of energy supply sources. In the Recommendation Concerning the Survey of Pumped-storage Electric Power Production, it called for hydropower surveys of pumped-storage power plant locations in order to establish a power source development approach that treats thermal, nuclear and pumped-storage power plants as harmonized sources.

Under these circumstances, the electric power companies also studied policies to promote hydropower development from a new perspective, thus establishing large-scale redevelopment plans for pumped-storage power plants, both stand-alone and as part of comprehensive development projects. Examples included the development of the Shin-Nariwagawa (PG, 103.0 m, #102) dam by Chugoku Electric Power Co., Inc. (ENERGIA), of the Azusa river comprehensive hydropower development project: Na-gawado, Midono, Inekoki and Takase (ER, 176.0 m, #63) dams by TEPCO, and of the Kuzuryu, Shin-Toyone and Tedorigawa (ER, 153.0 m, #60) dams by J-POWER.

A pumped-storage power plant requires two reservoirs: an upper and a lower reservoir. Until about 1970, many were constructed as mixed pumped-storage power plants[29] that could also produce ordinary hydropower where the inflow of river water to the upper reservoir was sufficient. However, as the number of available economical locations declined, the development of pure pumped-storage power plants at locations where either no water or extremely little water flows into the upper reservoir began to flourish, beginning with the station at Numappara (J-POWER, 1973) in the early 1970s. This was made possible by the development of new technologies: a steel penstock with a head drop in the 500 m class and high capacity reversible pump turbines.

Many efforts to reduce energy dependency on petroleum were initiated following the first and second oil shocks in 1973 and 1979, then in 1980, the Act on the Promotion of Development and Introduction of Alternative Energy was enacted, shifting priority to the construction of nuclear power plants.

Because both total electric power demand and the annual maximum demand stopped rising, almost all plans for new pumped-storage power plants have either been postponed or cancelled since the 1990s.

2.5.3.3.2   Pumped-storage hydropower plants and dams

Output from pumped-storage hydropower plants at the end of 1960 was only 58 MW (0.3% of total electric power production output), but it had grown to 3,390 MW (5.8% of total electric power output) by 1970.

Pumped-storage power plant output increased by about 20,000 MW from 1970 to 2001, as its share of all power production facilities rose from about 6% to 11% (Figure 2.5.15).

Table 2.5.3 shows examples of dams for pumped-storage power plant constructed since 1975.

The Takase Dam (Figure 2.5.16) is, with a dam height of 176 m, the highest rock-fill dam in Japan. Its construction has been described in the novel, Birth of Lake Water (Chuokoronsha, 1985), by Sono Ayako. The author collected information about changes that took place at the dam site to portray the entire world of dam construction in an extraordinary novel. This work was an inquiry about the meaning of civil works projects and the philosophical and religious significance at the root of civil engineering technologies, directed at the world at large.

The structure of power supply by power source is shown in Figure 2.5.17, revealing that in recent years, the base-load supply has been provided by run-of-river hydropower, nuclear power and coal-fired thermal power, load fluctuations during the daytime are handled by LNG and by LPG thermal power plants, and short-term peaks are supplied by dam-type hydropower and pumped-storage power.

---

29 Mixed pumped-storage power plants: in contrast to a pure pumped-storage type facility that uses electric power late at night to lift water from a lower reservoir to an upper reservoir, then during the daytime peak, releases this water to produce power, a mixed pumped-storage type combines normal hydropower production by the natural flow of the river into the upper reservoir with the pure pumped-storage method.

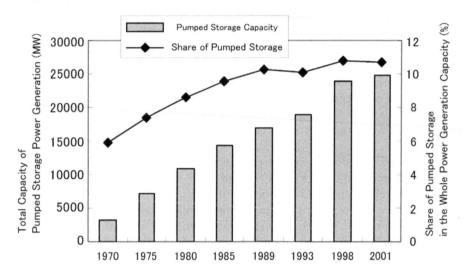

Figure 2.5.15 Change in the capacity of pumped storage power generation.

Source: METI [91].

2.5.4   Flood control by dams, based on the River Law

2.5.4.1   The Revised River Law (1964) and dams

The Previous River Law, enacted in 1896, focused on flood control and established a system in which river managers were prefectural governors or mayors of cities, towns, or villages, and only important or large-scale projects could be implemented directly by the central government (former Ministry of the Interior). As a result, wide-area management and efficient water management were hindered in a variety of ways, requiring radical revision.

Under these circumstances, the River Law was revised comprehensively in 1964 (referred to below as the "New River Law"). The New River Law classifies rivers as Class A Rivers or Class B Rivers, with Class A Rivers managed directly by the central government acting as river manager, guided by the concept of an integrated management of river system. Class B Rivers are managed by prefectural governors acting as river managers. It also established many new river management-related systems, including the authorization of water rights[30] by river managers, the enactment of dam operating regulations[31], instructions for action during floods and intake regulation for droughts, accompanied by special provisions concerning dams.

---

30 Water rights: rights to use water. The approval for the rights to exclusive use of water flowing in a river under Article 23 of the River Law are called authorized water rights and those approved historically and socially prior to the enactment of the law are called customary water rights.

31 Operating regulations: concretely specify the operating method when a structure is in use in order to achieve the desired purpose of the structure.

2.5.4.2 Roles of flood control dams in flood control plans and the effects of flood control

Flood control by dams can be counted on to have a faster effect than the widening of a river course that is not effective until a series of continuous river sections have been completed. Therefore, these have been planned to deal with massive floods that have occurred in succession: the West Japan Floods (1953), Kano River Typhoon (Typhoon Ida) (1958) and Ise Bay Typhoon (Super Typhoon Vera) (1959).

An examination of the Basic Plans for the Implementation of Construction Works stipulated by the New River Law (1964) reveals that of 109 Class A River Systems, as of 2004, 86 river systems were planned, including flood control by dams. The way that flood discharge is divided into discharge which is controlled by dams and that which is allowed to flow down the river course is established by the river manager, taking into consideration all conditions on each river system. The average allotment implemented on 86 river systems is, as shown in Figure 2.5.18, flood control of about 22% of the approximately 8,000 m^3/s unregulated peak design flood[32,33] at the downstream control point (Table 2.5.4).

As shown in Table 2.5.5, as of 2004, a total of 447 dams with flood control among their purposes (excluding farmland disaster protection dams) had been completed, ensuring approximately 4.4 billion m^3 of flood control capacity.

The Program Evaluation (policy review) of dam projects conducted in March 2003 by the MLIT has shown the following flood damage reduction effectiveness of dams [96].

"In Japan, flood control is performed to reduce flood damage for most of each year, mainly during the snow-melting season, the early summer rainy season and the typhoon season. An examination of the achievements of the 406 dams under the management of the MLIT (including dams managed by prefectures) shows that flood control was performed about 4,000 times during the ten-year period from 1991 to 2000. A total of 93 dams have been constructed, either by the central government or the JWA, to regulate floods, under the control of the MLIT. In comparison with about 3.7 trillion yen (2001 prices) which is the total investment in these projects (the flood control portion), it is estimated that the monetary value of the flood damage reduction effects of these dams was more than 4.2 trillion yen (2001 prices) during the 15 years from 1987 to 2001."

In this way, flood control by dams achieves large flood damage reduction effects and provides supplementary benefits. By capturing woody debris being carried downstream, it prevents blockage of the river at bridge piers, also preventing the inundation that this causes, resulting in a lower burden of flood-fighting activities required to protect levees by reducing the flood discharge level.

---

32 Unregulated peak design flood discharge: a flood that passes through a control point on the river completely unregulated by storage facilities is called the unregulated design flood, and its peak discharge is called the unregulated peak design flood discharge.

33 Design flood discharge: in contrast to the unregulated peak design flood discharge, the peak flood discharge that passes through a control point after flood control by dams is called the design flood discharge.

Table 2.5.3 Dams for pumped-storage power plant.

Name of power plant	Output (MW)	River system	Effective head (m)	Maximum water discharge (m³/s)	Upper pond				Lower pond				Commissioning year	Region	Owner
					Name of dam	Type	Height (m)	Length (m)	Name of dam	Type	Height (m)	Length (m)			
Niikappu	200	Niikappu	99.62	234	Niikappu	ER	102.8	326	Simonii-kappu	PG	46	131	1974	1	1
Takami	200	Shizunai	104.5	115	Takami	ER	120	435	Shizunai	PG	66	207.5	1983	1	1
Kyogoku	600	Shiribetsu	369	190.5	–	AFRD	22.6	1,108.6	Kyogoku	ER	54	332.5	(2015)	1	1
No. 2 Numazawa	460	Agano	214	250	Lake Numazawa	PG	53	168	Miyashita	PG	53	168	1982	2	2
Okukiyotsu	1,000	Shinano	470	260	Kassa	ER	90	487	Futai	ER	87	280	1978	4	10
Okukiyotsu No. 2	600			154									1996		10
Shimogo	1,000	Agano	387	314	Ouchi	ER	102	340	Okawa	PG	75	406.5	1988	2	10
Numappara	675	Naka	478	172.5	Numappara	AFRD	38	1,597	Miyama	AFRD	75.5	333.3	1973	3	10
Shiroyama	250	Sagami	153	192	Honzawa	ER	73	234	Shiroyama	PG	75	260	1965	3	11
Yagisawa	240	Tone	93.5	300	Yagisawa	VA	131	352	Sudagai	PG	72	194.4	1965	3	3
Tambara	1,200	Tone	518	276	Tambara	ER	116	570.1	Fujiwara	PG	95	230	1981	3	3
Imaichi	1,050	Tone	524	240	Kuriyama	ER	97.5	340	Imaichi	PG	75.5	177	1988	3	3
Shiobara	900	Naka	338	324	Yashio	AFRD	90.5	263	Sabigawa	PG	104	273	1994	3	3
Kazunogawa	1,600	Sagami	714	280	Kami-Hikawa	ER	87	494	Kazunogawa	PG	105.2	263.5	1999	3	3

Kannagawa	2,820	Tone	675	510	Minami-Aiki	ER	136	444	Ueno	PG	120	350	2005	3	3
Azumi	623	Shinano	135.8	540	Nagawado	VA	155	355.5	Midono	VA	95.5	343.3	1969	5	3
Midono	245	Shiano	79.8	360	Midono	VA	95.5	343.3	Inekoki	VA	60	192.8	1969	5	3
Shin-Takas-egawa	1,280	Shinano	229	644	Takase	ER	176	362	Nanakura	ER	125	340	1979	5	3
Hatanagi No.1	137	Oi	101.2	160	Hatanagi No.1	PG	125	292	Hatanagi No.2	PG	69	171	1962	5	5
Takane No.1	340	Kiso	135	300	Takane No.1	VA	133	276.4	Takane No.2	PG	69	232	1969	5	5
Mazegawa No.1	288	Kiso	99.6	335	Iwaya	ER	127.5	366	Mazegawa No.2	PG	44.5	263	1976	5	5
Oku-Yahagi No.1	315	Yahagi	161.3	234	Kuroda	PG	45.2	332	Tominaga	PG	32.5	337	1980	5	5
Oku-Yahagi No.2	780	Yahagi	404.6	234	Tominaga	PG	32.5	337	Yahagi	VA	100	323.1	1980	5	5
Oku-Mino	1,500	Kiso	485.8	375	Kaore	VA	107.5	341.2	Kami-Osu	ER	989	294.5	1994	5	5
Shin-Toyone	1,125	Tenryu	203	645	Shin-Toyone	VA	116.5	311	Sakuma	PG	155.5	293.5	1972	5	5
Nagano	220	Kuzuryu	97.5	265	Kuzuryu	ER	128	355	Washi	VA	44	277	1968	4	10
Ikehara	350	Kumano	120.5	342	Ikehara	VA	111	459	Nanairo	VA	61	200.8	1964	6	10
Kisen-yama	466	Yodo	219.4	248	Kisen-yama	ER	91	255	Amagase	VA	73	254	1970	6	6
Oku-Tataragi	1,930	Ichi	383.4	376	Kurokawa	ER	98	325	Tataragi	AFRD	64.5	278	1974	6	6
Oku-Yoshino	1,206	Shingu	505	288	Seto	ER	110.5	342.8	Asahi	VA	86.1	199.4	1980	6	6
Ohkawachi	1,280	Ichi	394.7	382	Ohta No.1	ER	55.5	175.3	Hase	PG	102	254	1992	6	6

(Continued)

Table 2.5.3 (Continued).

Name of power plant	Output (MW)	River system	Effective head (m)	Maximum water discharge (m³/s)	Upper pond Name of dam	Type	Height (m)	Length (m)	Lower pond Name of dam	Type	Height (m)	Length (m)	Commissioning year	Region	Owner
Shin-Nari-wagawa	303	Takahashi	84.7	424	Shin-Nari-wagawa	PG	103	289	Tabara	PG	41	206	1968	7	7
Nabara	620	Ota	294	254	Myoji	ER	88.5	402	Nabara	ER	85.5	305	1976	7	7
Matanogawa	1,200	Hino	489	300	Doyo	ER	86.7	480	Mata-nogawa	PG	69.3	185	1986	7	7
Honkawa	615	Yoshino	528.4	140	Inamura	ER	88	352	Ohashi	PG	73.5	187.1	1982	8	8
Odaira	500	Kuma	490	124	Uchitani	ER	64	200	Aburatani	ER	82	189.2	1975	9	9
Tenzan	600	Matsuura	520	140	Tenzan	ER	69	380	Itsuki	PG	117	390.4	1986	9	9
Omarugawa	1,200	Omaru	646.2	222	Ohseuchi	AFRD	65.5	166	Ishika-wauchi	PG	47.5	185	2007	9	9
					Kanasumi	AFRD	42.5	140							

Remarks

Number	Owner
1	Hokkaido Electric Power Company Inc.
2	Tohoku Electric Power Company Inc.
3	Tokyo Electric Power Company Inc.
4	Hokuriku Electric Power Company Inc.
5	Chubu Electric Power Company Inc.
6	Kansai Electric Power Company Inc.
7	Chugoku Electric Power Company Inc.
8	Shikoku Electric Power Company Inc.
9	Kyushu Electric Power Company Inc.
10	Electric Power Development Co., Ltd.
11	Kanagawa Pref.

Region	
Hokkaido	
Tohoku	
Kanto	
Hokuriku	
Chubu	
Kansai	
Chugoku	
Shikoku	
Kyushu	

Presented by TEPCO

Figure 2.5.16 Takase Dam. (See colour plate section)

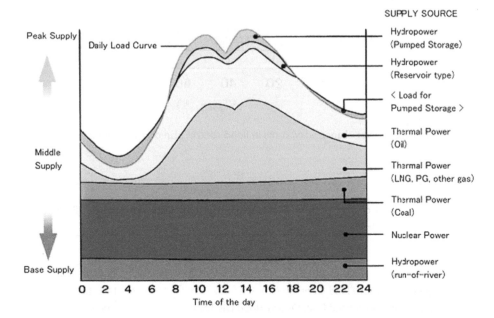

Figure 2.5.17 Combinations of electric power supplies by the time of day.
Source: FEPC [92].

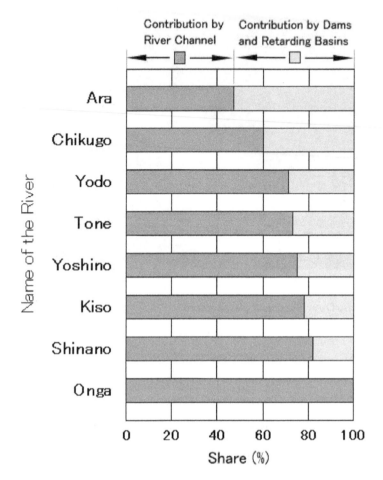

Figure 2.5.18 Share of river channels and dams in flood control plan of major rivers.
Source: MLIT.

Table 2.5.4 Flood control plan for Class A River Systems with dam projects (86 River Systems).

Number of Dams	256
Total Capacity for Flood Control	3,606 million m³
Average Capacity for Flood Control	14.1 million m³
a Average of Unregulated Peak Design Flood Discharge	8,000 m³/s
b Average of Peak Design Flood Discharge	6,200 m³/s
c b/a	78%

Source: Japan River [94], Japan Dam Foundation [95].

Table 2.5.5 Number of dams constructed and flood control capacity.

Year of completion	Number of dams constructed			Flood control capacity (million m³)		
	Flood control purpose only	Multi-purpose	Total	Flood control purpose only	Multi-purpose	Total
1926–1945	0	2	2	0	10	10
1946–1955	0	17	17	0	240	240
1956–1965	2	50	52	1	770	771
1966–1975	21	62	83	54	860	914
1976–1985	26	76	102	67	849	916
1986–1995	13	80	93	34	709	743
1996–2004	14	84	98	21	747	768
Total	76	371	447	177	4,185	4,362

Source: Japan Dam Foundation [95].

In this way, flood control by dams, with the help of downstream river improvements, has effectively reduced flood damage on the alluvial plains where assets are concentrated.

2.5.4.3   Effectiveness of the flood control by dams in recent years

Typhoon No.16 that hit Kagoshima Prefecture in Kyushu Island on August 30, 2004 traversed the Japanese Archipelago, causing severe damage along the Yoshino River in Shikoku Island where it flooded 57 houses and submerged 200 hectares of land.

The runoff triggered by this typhoon, a flood with a peak discharge of 4,000 m³/s or the equivalent of 85% of the designed flood discharge of 4,700 m³/s, flowed into the Sameura Dam (PG, 106.0 m, #117, Figure 2.5.19, Figure 2.5.20). The Sameura Dam's flood-control capacity (90 million m³) was used to store about 54 million m³ of flood in the dam, reducing the quantity discharged to a maximum of 1,774 m³/s, and with the help of the Dozan River Dam Group (the Shingu Dam, the Yanase Dam (#112), and the Tomisato Dam), and Ikeda Dam, it lowered the water level by 1.0 m near the downstream Miyoshi Bridge, thereby reducing flood damage [97].

2.5.4.4   Limitations on regulation functions of flood control dams [98, 99]

A dam regulates floods by storing part or all of the flood discharge water entering its reservoir, thus reducing the quantity of water that is allowed to continue flowing downstream. Each dam has flood control functions according to its reservoir storage capacity, but this capacity is subject to limitations. At each dam, the quantity of floodwater that is the object of the flood control plan is set in advance. The part of the flood discharge to be stored, the part to be allowed to flow on without being

Presented by JWA

Figure 2.5.19 Sameura Dam. (See colour plate section)

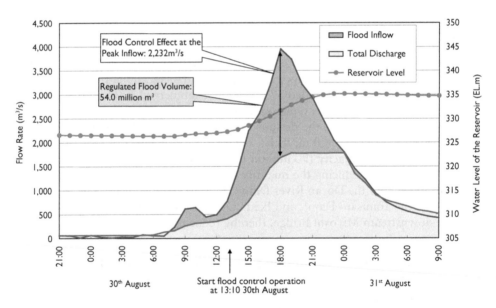

Figure 2.5.20 Flood control operation of the Sameura Dam during Typhoon No. 16 in 2004.
Source: JWA.

stored, and the way that stored water is discharged later, are all determined according to operating regulations. If the reservoir is filled as a result of the flood control, based on the operating regulations, it cannot store any more floodwater, or in brief, cannot perform further flood control. In this case, the manager switches to crisis management to deal with the fury of nature, which has exceeded the capacity of the facility and the plan.

The limitations of flood control functions have still not been adequately explained, and it is assumed that they are, therefore, not fully understood by the residents occupying the land along downstream reach of rivers.

The flood control capacity of each dam is usually set with about 20% leeway, but when a flood is so large that it exceeds the scale of the planned flood (this is called an "exceeding flood"), it may exceed the dam's flood-control capacity, so the moment that it is clear that this has occurred, a special operation begins that allows the inflowing flood to continue flowing downstream in an unregulated manner. At a dam constructed to control floods, the limitations on the dam's flood control capacity must be explained to residents occupying the land along the river downstream from the dam, and according to circumstances, the concept of the flood control plan itself must be revised.

Major instances of special operations that were undertaken when an exceeding flood reached the limitation of the flood control capacity of a dam occurred at the Yahagi Dam (VA, 100.0 m, #84) on the Yahagi River during the flood discharge triggered by torrential rainfall in September 2000, at the Nibutani Dam (PG, 32.0 m, #6) on the Saru River during a flood caused by Typhoon No. 10 in August 2003, and the Nomura Dam (PG, 60.0 m, #116) on the Hiji River during a flood caused by Typhoon No. 16 in August 2004.

As shown by the above account, dams have been constructed and operated in order to provide irrigation water or municipal and industrial water, to generate electric power and to control floods, according to the needs of each era in each region of Japan. This is one important element in the way that the people have worked to expand land that is appropriate for production and increase the size of areas where people can live safely on the archipelago of Japan. As a result, river development at every stage of history has laid down a historically-layered structure to form the present water cycle on the national land of Japan and living organisms of many kinds continue to inhabit this water cycle. This endless development of national land has enabled us to enjoy the blessings it has brought us: increasing our productive capacity and enabling us to live in safety.

Chapter 3

Environmental and social impacts of dams and responses to these impacts

As shown in Chapter 2, dams have been constructed in Japan to supply water or to control floods, creating benefits that are difficult to obtain by other means. However, as large-scale dams were constructed by using modern technology in various regions since the Meiji Period, it became impossible to ignore environmental and social impacts of dams.

This impact first surfaced in the form of public disputes over the expropriation of land for dam projects. Reflection on this situation resulted in the enactment of laws that established a system of providing compensation for losses caused by dam projects and provided support for the mitigation of the widespread impact of projects on reservoir areas and for the activation of these regions. Not only the direct impact of submersion, but change in downstream flow regimes from the construction of dam reservoirs, the progress of sedimentation upstream from reservoirs, the discharge of cold water and turbid water, etc., have a variety of impacts on the use of river water or habitats of living organisms, thus creating many social problems. It also established a framework for research, development and implementation of technological measures to mitigate these impacts, and for environmental impact assessment to conserve the environment.

In response to intensifying discussions on the appropriateness of dam projects as public work projects and the accountability for these projects, in recent years, measures to reflect citizens' opinions of dam projects have been introduced, together with public project assessment systems that are necessary to implement, cancel or continue projects. These trends are explained below.

3.1 DAM PROJECTS AND SOCIAL CONFLICT

The impacts of dam projects on social and economic conditions and on the natural environment create tense relationships between organizations that implement projects and local residents, and this tension has often been expressed in rising social conflicts. Such conflicts occur in various forms including litigation, opposition movements involving the media, the submissions of petitions to and questions in the Diet, and so on.

Social conflicts surrounding dam projects began with disputes regarding the construction of the Shimouke and Matsubara Dams in Kyushu Island. In 1953, on the Chikugo River, an unprecedented flood disaster killed 147 people and submerged 67,000 hectares of paddy and dry fields. In response, the MLIT enacted the Flood Control Basic Plan for the Chikugo River, which included plans for the construction of the Shimouke Dam (VA, 98.0 m, #128) and Matsubara Dam (PG, 83.0 m, #127) in the upstream region of the Chikugo River.

At the Shimouke Dam, an historical anti-dam movement called the Hachinosu Castle Conflict prevailed from 1955 to 1970. It is reported that major reasons for the intensification of this anti-dam movement led by Murohara Tomoyuki (a forest owner with rights to the Shimouke Dam site) were the government changing the original locations selected for the dams to the Shimouke and Matsubara sites further upstream without giving sufficient explanation on reasons for the change to the people affected by the projects, and for its failure to explain to the landowners how they would be resettled at the meetings.

While Murohara Tomoyuki refused to negotiate with the MLIT, he led a series of 80 lawsuits, driven by the slogan, "Fight violence with violence, law with law" and the ideal of, "Public works projects must comply with reason, comply with law, and comply with human feelings".

In 1960, he started his key attack: an administrative lawsuit demanding that the project's approval under the Land Expropriation Law be nullified, but this lawsuit was decided in favor of the central government in 1963. Opponents of the project constantly manned the Hachinosu Castle constructed on the right riverbank at the Shimouke Dam site, preventing officials of the MLIT from entering the site. Hachinosu Castle was forcefully closed in 1964 and then, for a second time in 1965, all buildings at the dam site were removed. After the death of Mr. Murohara in 1970, peace between his relatives and the MLIT returned after 16 years, thus ending the dam opposition movement.

This conflict, which was the first opposition movement that disrupted the sacred realm called the "public interest", had a major impact on the 1964 revision of the River Law and the government's future conduct regarding public projects.

From 1972 to 1975, other major lawsuits were launched regarding the construction and the management of dams: the Kotogawa Dam (PG, 38.8 m, #106), the Tsuruta Dam (PG, 117.5 m, #133), the Shin-Nariwagawa Dam and the Nagayasuguchi Dam (PG, 85.5 m, #109). All were a result of flood disasters in 1971 and in 1972. During this period, the Kotogawa Dam and the Shin-Nariwagawa Dam cases were appealed to the court of second instance where the plaintiff discontinued the action and accepted a compromise settlement, but the Tsuruta Dam and Nagayasuguchi Dam lawsuits reached the Supreme Court.

As for the Tsuruta Dam case, on July 6, 1972, concentrated torrential rainfall triggered a flood, washing away houses and submerging others to floor level along the Sendai River, downstream from the dam. A group of 123 residents who suffered from this disaster sought compensation for losses from the central government, claiming that the disaster was caused by insufficient flood control capacity at the Tsuruta Dam and by a massive discharge from this dam. In both the first and the second level courts, the defendant (the central government) won, and the government's victory was confirmed by the Supreme Court in 1993.

At the Tsuruta Dam, the flood control capacity was increased from 42 million m^3 to 75 million m^3 in June 1973.

The public demanded more thorough management in dealing with the abrupt rise in the downstream river water level that occurs when discharge begins, and in providing adequate advance warning to the public when starting discharge from the dam: two issues that were raised in the litigation.

These conflicts were bitter experiences for all the dam managing bodies and triggered later response measures.

Beginning in part 3.2, the impacts of dams and measures to mitigate or respond to such impacts are divided into four fields, including instances that did not always lead to litigation, to clarify recent trends. In the following section, the social impact on reservoir areas and responses to deal with such impact are examined.

3.2   IMPACTS ON THE SOCIAL CONDITION IN RESERVOIR AREAS AND RESPONSES TO SUCH IMPACTS

3.2.1   The social impact of dam projects on reservoir areas and responses to such impact

Many experts have pointed out that while the construction of dams provides great benefits, mainly for downstream beneficiary regions or the people's economic life, they also impact negatively on the economies and public life of regions submerged by the dam, and on surrounding regions.

Not only do they submerge privately-owned and public assets, but the submerging also destroys or shrinks hamlets, divides the region, and has many other negative effects.

Since the end of World War II, mountainous regions have continually faced a number of problems. In their roles as residential regions, an aging population and depopulation, and as areas for economic and productive activities, the decline of agricultural, forestry and other local industries. Through the rapid growth of Japan's economy and depopulation of the mountain villages, the impacts of dam projects were not always clearly distinguished from these other trends, but there have been instances where the start of a dam project abruptly revealed these problems. Therefore, through the efforts of stakeholders, a response based on reservoir area development measures with the configuration shown in Table 3.2.1, has begun. This section describes these measures and the history of their expansion.

3.2.2   Compensation measures by dam project organizations for residents submerged by dam projects and for related municipalities

Residents submerged by the construction of a dam lose their homes, their land and their occupations, which are the foundations of their livelihoods and daily lives, so to mitigate these impacts, it is necessary to take so-called measures for resettlement of the inhabitants. The core of these livelihood reconstruction measures is the payment of compensation by the project organizations.

Table 3.2.1 Structure of reservoir area development measures.

Reservoir area development measures
    Livelihood reconstruction measures for the relocated residents
        Securement of site and housing
            General compensation (Compensation in money for housing and other private
                properties)
            Preparation of substitute housing
            Preparation of substitute farm land
        Securement of employment opportunity
            General compensation (Compensation in money for discontinuation of business etc.)
            Go-between activities for employment of the relocated residents
        Support for stable livelihood
            Consulation for livelihood reconstruction
            Offering loan and tax incentives
    Improvement measures for the reservoir area
        Securement of public utility functions
            Public compensation (Compensation or providing substitute facilities; eg. road,
                school, etc.)
        Support for the stability of the area
            Provision of social infrastructure
            Conservation of local culture
            Preferential treatment for the finance of municipalities
        Support for economic enhancement of the area
            Support for industrial and agricultural development
            Creation of tourism resources
            Promotion of inter-regional exchange

Source: Water Resources Environment Technology Center [100].

During the 1950s, Japan achieved astonishing economic growth. This was accompanied by work to improve roads, rivers and so on in order to expand public infrastructure that was in short supply at the time, but it became increasingly difficult to deal with the problem of compensation necessary to secure the land needed for each project, making this problem the focus of public attention. Without uniform standards stipulating the objects of compensation being agreed upon by bodies implementing each project and, details of the compensation system and methods of providing it, instances of inappropriate compensation were sometimes discovered.

This was a period of expansion of many public works projects in addition to dam project; a time when dam project organizations strived to devise ways of conducting projects effectively, and the government eagerly tried to devise the public lands compensation system as explained below.

The enforcement of the new Land Expropriation Act in 1951 clarified the rights covered by compensation and firmly established the concept (principles) of compensation for these rights. In 1953, the Cabinet agreed on the Rules for Compensation for Loss by Submergence of Land and Other Impacts of Electric Power Development, and this was followed by the stipulation of similar rules for projects of the MLIT and land improvement projects. However, to unify these standards, which varied between project organizations, and to provide compensation more equitably and appropriately, the Cabinet promulgated the Guideline of Compensation for Loss Accompanying the Expropriation of Public Land in 1962. This is called the General Compensation Standard, because it presents standards of compensation for loss resulting from the acquisition or use of land required for public works projects. In 1967, the Cabinet enacted the Guideline of Compensation Standard for Public Property. This document, which provides standards for compensation paid for the loss of the functionality of a public facility caused by a public works project – the submersion of an existing road by a dam for example – is called the Public Compensation Standard. These compensation guidelines clarified the range of payment of compensation and organized and unified compensation items, so that compensation for each item could be calculated with a unified method.

The General Compensation Standard, in principle, limits the objects of compensation to property rights, but does not cover the right to life or spiritual damage. However, at many dams, compensation for remaining minorities[1] or job separation compensation systems have been established. Regarding measures to resettle residents of submerged land, in the Enforcing Guidelines for Compensation for Loss by Expropriation of Public Property (approved by the Cabinet in 1962) that were enacted at the same time as the Guidelines, it is stipulated that, "When there are people who have lost the foundations of their lives as a consequence of the execution of a project for the public benefit, as necessary, the project organization will strive to act as mediator to provide them with new land or buildings so they can relocate their homes and to take measures to provide introductions and guidance so they can obtain new employment."

Nevertheless, measures to relocate recipients of compensation are far more dependent on efforts by dam project organizations and relevant local government bodies, than on legal provisions. By the 1950s, dam project organizations were already preparing farm and housing land (the Sannokai, Ikawa, Miwa, Matsubara and Shimouke Dams), preparing land for group resettlement (the Yuda and Yanase (ER, 115.0 m, #118) Dams), acting as go-betweens to obtain farmland (the Sakuma, Makio and Shijushida Dams), finding employment and introducing vocational training opportunities.

At the Ikawa Dam, the residents of the submerged region moved to new residential land and public facilities were gathered in the center of the town, based on the concept of town creation through mediation by the prefecture. Farmland was prepared on the Nishiyamadaira Plain, to be used for rice cultivation, which had previously been impossible. Compensation for relocation was paid to 194 households and the creation of a new town in this instance later became a model for hydropower development throughout Japan. Furthermore, at the Shiroyama Dam (PG, 75.0 m, #50, 280 houses relocated) on the Sagami River System, the prefecture took a wide range of measures, such as: preparing group relocation land[2], acting as mediator to obtain

---

1 Compensation for remaining minorities: this is paid to a small number of people who are left behind in a hamlet when most other residents have moved because their land has been expropriated for a public works project, and are in a situation where their hamlet has lost its functionality as an economically collective body in society.
2 Group relocation land: new replacement land that is prepared as a relocation site for a hamlet that has been submerged because of the construction of a dam.

the desired land, offering priority admission to public housing, assisting relocated workers find new employment, offering vocational training, arranging consultations concerning commerce and industry, providing funds, and reducing and exempting real-estate acquisition tax rates.

In 1979, the cost of the measures for resettlement of the inhabitants of the reservoir area, and in 1983, the cost of restoring the living environment, were included in dam construction costs of projects undertaken directly by the MLIT, permitting even more thorough measures to resettle inhabitants and restore their normal daily lives [101].

Examples of public compensation range widely, including the diversion of railway tracks (Sakuma Dam, Yuda Dam), diversion of roads, relocation of elementary and junior high schools and city halls, and installation of waterworks in the relocation areas, etc.

To compensate for business losses caused by the implementation of a dam project that cannot be adequately compensated for under General and Public Compensation Standards because their impact extends to an unspecified large number of people in the project region, related public projects have been initiated in order to mitigate the impact of such losses (Makio Dam: a prefectural road, forest road, etc.).

3.2.3   Support for submerge-related municipalities by central and regional governments, etc.

Dam projects have a strong impact on submerged land, so in Japan, not only compensation by dam project organizations, but also various support measures have been established for local government bodies including submerged regions. The following is an outline of these, with the focus on central government measures, based on special laws or measures in the budget system and on support by downstream beneficiary regions.

3.2.3.1   Regional development based on the Law Concerning Special Measures for Reservoir Area Development, in respect of designated dams

When the impact of a dam project on a region is particularly great – many homes are submerged for example – simply relocating the residents with private and public compensation is not adequate. During the 1960s (the latter half), the construction of dams that submerged more than 200 homes or households began to appear: the Gosho Dam (submerged 448 homes), the Yamba Dam (340 households), the Miyagase Dam (PG, 156.0 m, #52, 281 homes), the Tedorigawa Dam (330 households), the Otaki Dam (PG, 100.0 m, #98, 399 homes), the Tomata Dam (PG, 74.0 m, #101, 504 households) and the Kawabegawa Dam (VA, 107.5 m, #131, 549 households). As a result, it became necessary to take wide ranging measures in upstream reservoir regions – resettling the residents affected by the submersion of the land, providing a living infrastructure in upstream reservoir regions and taking action to strengthen mutual understanding between upstream and downstream residents – mainly by compensation from the dam project organization [102–106].

Public bodies in upstream reservoir regions had to provide infrastructure that was larger than normal to mitigate the impacts of the construction of a dam. It was necessary to increase the central government's subsidy rate for infrastructure development projects and for local government bodies in downstream beneficiary regions to begin to bear part of the costs in order to carry out dam projects effectively.

To deal with these circumstances, the Law Concerning Special Measures for Reservoir Area Development (called "RAD law" below) was proclaimed in 1973. The purpose of this law is to stabilize the concerned residents' lives, improve their welfare and encourage the construction of dams by taking a variety of measures to mitigate the impact on the productivity and living environment of regions severely impacted by dams or lake or marsh water level control structures. In 1994, preventing the water pollution in dam reservoirs was added, and revisions were made to expand measures to revitalize reservoir areas.

As of 2007, for submersion by a dam to be subject to the RAD law, it must result in 20 or more submerged homes and 20 hectares or more of submerged farmland (60 hectares or more in Hokkaido). After designation, the prefectural governor prepares a plan for a reservoir area development project, and the Minister of Land, Infrastructure, Transport and Tourism makes a final decision on the reservoir area development plan.

During the period from the enactment of the RAD law in 1974 to the end of 1999, 94 dams and one lake were designated.

The RAD law clearly stipulates 8 projects, including land improvement projects, that are considered to be reservoir area development projects and another 16 types of projects are prescribed in the Enforcement Ordinance to the RAD law. Table 3.2.2 shows examples of the law's application to major dams.

An examination of the contents of development projects implemented in 1998 at these dams reveals that, as shown by Figure 3.2.1, about 67% of them are road, land improvement and flood control projects that are related public facilities, and that the

Table 3.2.2 Examples of application to major dams of the Law Concerning Special Measures for Reservoir Area Development.

Name of dam	Implementing organization	Location (Prefecture)	Submersion by dam Area (ha)	Number of houses	Farmland (ha)	Year of commencement	Dam project cost (billion yen)	Cost for reservoir area activation measures (billion yen)
Aseishi-gawa (PG, 91.0 m, #9)	MLIT	Aomori	222	201	59	1975	90.5	3.761
Gosho	MLIT	Iwate	640	440	390	1975	48.88	11.984
Miyagase	MLIT	Kanagawa	490	300	19	1980	397.0	29.158
Tedori-gawa	MLIT J-POWER Ishikawa Pref.	Ishikawa	525	330	33	1975	74.0	16.966
Otaki	MLIT	Nara	240	399	8	1979	298.0	12.001

Source: Water Resources Environment Technology Center [107].

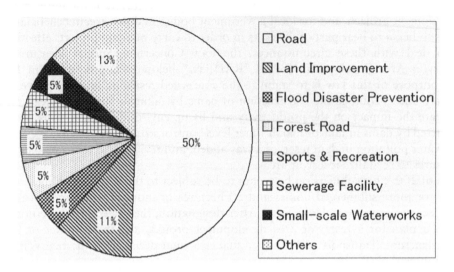

Figure 3.2.1 Breakdown of costs for reservoir area development measures by type of project (1998).
Source: Water Resources Environment Technology Center [108].

burden of paying the cost of these projects was: the central government approx. 47%; prefectures approx. 23%; submerge-related municipalities about 27%; and others about 2% [108].

The following is an example of sharing of the cost of reservoir area measures, based on the RAD law at the Tedorigawa Dam constructed by raising the central government's share of costs (MLIT, 1980). Table 3.2.3 is a comprehensive table of development work under the RAD law at the Tedorigawa Dam. The share borne by submerge-related municipalities was 510 million yen, or 3.0% of the total development project cost of 17 billion yen [109].

The Tedorigawa Dam was designated under the Act on the Development of Areas Adjacent to Electric Power Generating Facilities and constructed in accordance with reservoir area development, based on the Three Power Source Laws explained below.

3.2.3.2 Support by the fund for reservoir area development

Funding systems that harmonize the benefits of water use between the upstream and downstream regions can be traced back to the system in Yokohama City, which started in 1916 when Kanagawa Prefecture sold the Imperial Forests. Since then, Yokohama City has borne part of the costs of equipment for public facilities in Doshi Village, which is its reservoir area, employs forestry workers, and conducts cooperative projects.

In 1969, Yamagata Prefecture began lending funds to residents of land submerged by the construction of the Shirakawa Dam (ER, 66.0 m, #20) to make an advance purchase of real estate for relocation housing. It also established a system to subsidize the interest they paid on loans obtained from financial institutions for the same

Table 3.2.3 Classification of projects and burdens under the Law Concerning Special Measures for Reservoir Area Development (Tedorigawa Dam).

Type of project	Total cost (million yen)	Cost allocation (million yen)			
		National government	Prefectural government	Municipal government	Beneficiary bodies
Land improvement	31	15	3	12	2
Forest conservation	1,138	853	284	–	–
Flood disaster prevention	4,900	3,367	1,533	–	–
Road	9,372	6,939	2,414	19	–
Small-scale water-works	306	64	–	242	–
Compulsory education facilities	452	219	–	233	–
Housing land development	342	–	342	–	–
Publicly-owned apartment	424	141	283	–	–
Total	16,966	11,398	4,860	506	2

Source: T. Maruyama, 1986 [109].

purpose. Between 1969 and 1976, 18 prefectures introduced similar interest subsidies for 28 dams.

Progress was achieved when the reservoir and beneficiary areas were under the same local government, but in instances of wide area water usage where the reservoir and beneficiary areas were in two or more prefectures, the system could not be introduced because of a restriction imposed by the Local Finance Law, and the difficulty of harmonizing interests resulted in strong demand from both residents of submerged lands and from local public bodies for the establishment of a new organization to deal with these problems. The Lake Biwa Management and Coordination Fund System, which was strengthened in conjunction with the Lake Biwa Comprehensive Development Project, was the first actual funding system for current wide area water usage.

In the above circumstances, in 1976 Tokyo and 5 prefectures with interests in the Tone and Ara River Systems, and other concerned bodies in the national government, initiated the Tone River – Ara River Reservoir Area Development Fund. Later, 7 funds, including the Kiso Three River Reservoir Area Development Fund, were established with subsidization by the central government as reservoir area development funds, and 25 funds were under operation as major reservoir area development funds as of fiscal 1998.

These funds subsidized the cost of projects, mainly to relocate residents, to obtain real estate, to stimulate and develop reservoir areas, and to hold events that promote close relationships between residents of the upstream and downstream regions of a river system, with relevant local government bodies bearing the cost. Upstream reservoir area development funds supplemented legal provisions, and there were instances where, after a dam was completed, projects continued with the support of these funds.

3.2.3.3   The system of subsidies for municipalities with dam sites
in their jurisdiction

In 1974, at about the same time that the RAD law was enacted, a system was established to subsidize municipalities that could not levy fixed property taxes on waterworks or industrial water dams constructed within their boundaries, to compensate the municipalities for the taxes they lost because of dams.

In this way, municipalities in which water use dams were located, excluding irrigation dams, received a subsidy income.

The total amount of subsidies paid for specified multi-purpose dams in 1998 amounted to 2,644 million yen for 38 dams, an average of about 70 million yen per dam. At these dams, the amounts varied according to the year of completion and the amount of the construction cost borne by water users [110].

3.2.3.4   Support for reservoir area municipalities
by electric power projects

During the 1970s, increasing difficulty in obtaining locations for electric power plants was accompanied by a series of oil crises and tight supply and demand for electricity. So in 1974, the so-called Three Power Source Laws, which included the Act on the Development of Areas Adjacent to Electric Power Generating Facilities, were enacted to support the acquisition of electric power plant locations. A system funded by electric power development promotion taxes paid by electric power companies was established to provide public facilities in the region surrounding electric power plant sites. This system, which also applies to hydropower generation facilities, completed the support systems for submerge-related municipalities.

Under the Act on the Development of Areas Adjacent to Electric Power Generating Facilities, power plant location promotion measure subsidies calculated according to the output of each electric power-generation facility are paid to municipalities that establish electric power generators and the municipalities apply these subsidies to public facility development in the area adjacent to the electric power-generating facility. The subsidies for hydropower plants are paid to municipalities where facilities with an output of 1,000 kW or more are constructed. The central government pays subsidies to the municipalities where power-generation facilities are located and to nearby municipalities based on a Development Plan prepared by a prefectural governor and approved by the responsible government minister.

Table 3.2.4 shows examples of major dams operated by the MLIT that have received location designations under the Act on the Development of Areas Adjacent to Electric Power Generating Facilities.

A total of approximately 6 billion yen in annual subsidies have been paid as hydropower plant adjacent region subsidies in recent years. At the Tedorigawa Dam, reservoir area development has been carried out under the RAD law. It has also been promoted under the designation of the Act on the Development of Areas Adjacent to Electric Power Generating Facilities. A total of 275 million yen has been paid in subsidies to the concerned municipalities. Power resource subsidies account for an average of over 50% of the cost of development projects by local municipalities [112].

Table 3.2.4 Examples of dams that have received location designations under the act on the development of areas adjacent to electric power generating facilities.

Name of dam	Implementation organization	Project cost for the outskirts area (million yen)	Subsidy from the national govern- ment (million yen)
Okawa (PG, 75.0 m, #22)	MLIT	615	610
Yahagi	MLIT	879	810
Tedorigawa	MLIT	514	276
Miyagase	MLIT	255	140
Jozankei	MLIT	334	45
Nishimura	Gifu Prefecture	194	175
Masaki	Tokushima Prefecture	827	74

Source: Water Resources Environment Technology Center [111].

3.2.4 Support for the activation of reservoir areas

In addition to financial support and facility construction by dam project organizations, the central government, concerned local governments and dam organizations have, in recent years, taken non-physical measures to activate reservoir areas by introducing measures such as making use of dam reservoirs and holding events to strengthen relationships between upstream and downstream area residents.

3.2.4.1 The use of the surroundings of dam reservoirs

The recreational use of the surroundings of dam reservoirs at new dams began in the late 1950s, mainly at electric power dams. This includes services such as PR buildings, rest facilities, the sale of traditional local products and the operation of pleasure boats on the reservoirs for tourists who come to enjoy the beauty of large dams and reservoir water. During the 1970s, at dams operated directly by the MLIT, parks, etc. were developed cooperatively with municipalities as projects to develop the environment surrounding dams. Then, in the late 1980s, projects such as dam reservoir utilization promotion projects were carried out at existing dams, thus activating reservoir regions.

3.2.4.2 Dams open to their surrounding communities [113]

In 1992, this trend was expanded under the slogan, "Dams open to their surrounding communities". This promoted the use of dams and reservoirs among residents of the surrounding areas by increasing the accessibility of dams, so that the facilities become more familiar to them. Dams where such measures are taken are designated on the premise that local residents will actively use the dam facilities, etc. By the end of fiscal 2003, high-quality memorial buildings, archives and other PR facilities, plus surrounding parkland had been provided at 44 dams. As these new facilities were provided, municipalities took their completion as opportunities to establish third- sector enterprises to partially manage the facilities, to provide locations for the study of nature and of human society, and to offer services to tourists.

At some dams, such facilities were also used to hold events to support the relocation of those displaced by the dams, so they play major roles in regional revitalization.

3.2.4.3   Measures to promote social interchange

The promotion of inter-regional exchanges – mainly between residents of upper and lower reaches of river basins – began in the 1970s in the Sagami and Yahagi River Basins when water was used as a medium for social interchange. In recent years, many dams have expanded the interchange between their downstream beneficiary areas and their reservoir areas.

Taking Chiba Prefecture as an example, it was believed that bringing people from the upstream region together with prefectural residents in the downstream basin who benefit from the dam would increase their mutual understanding of each group's actual circumstances, and nurture a consciousness of their solidarity and finally lead to rapid completion of the water resources development facilities. Beginning in fiscal 1992, Chiba Prefecture initiated its own prefectural project to foster closer relationships between upstream and downstream residents by holding events with a central theme of "water", which involved Naganohara Town in Gunma Prefecture and Kuriyama Town in Tochigi Prefecture: the planned construction sites of the Yamba and Yunishigawa (PG, 130.0 m, #23) Dams respectively [114].

Then, in 1977, the cabinet approved the designation of August 1 as Water Day. The week beginning August 1 was declared Water Week: the week when, throughout Japan, events including exchanges between residents of reservoir areas were held to publicize the importance of water.

Then in 1987, the MLIT and the Forestry Agency, MAFF, established Ten Days to Get to Know Our Forests and Reservoirs (July 21 to July 31) as a joint project intended to give the people of Japan an opportunity to get acquainted with their forests and reservoirs in order to refresh themselves physically and spiritually and nurture their energy for future endeavors, and at the same time heighten their concern for, and deepen their understanding of, the importance of dams, forests and other water resources [115]. During that 10-day period, events planned to introduce people to forests and reservoirs were held simultaneously at dams and forests throughout Japan along with art contests with the theme of forests and reservoirs and tours of reservoir areas, thus contributing to the establishment of good relations between upstream and downstream communities.

3.2.4.4   Future measures in reservoir areas

In many reservoir areas, depopulation caused by the outflow of people, the aging of society, and the weakening of industrial infrastructure are hindering regional development. To address these problems, the then National Land Agency began the reservoir area measure adviser dispatch system in 1988 to assist with reservoir area activation along with dam construction. Under this system, guidance teams made up of three or four advisors are dispatched to each area in response to a request from a reservoir area municipality. They have visited a total of 31 areas.

Regardless of these initiatives, many reservoir areas are still in need of activation. So in 2001, the MLIT enacted a Reservoir Area Vision for each dam operated by it and the JWA as a 21st century reservoir area measure [116]. These are action plans intended to utilize dams to promote independent and sustained activation of reservoir areas and to achieve balanced development of their river basins, by establishing links and carrying out exchange activities between people throughout each basin. These action plans are led by municipalities and residents of dam reservoir areas working cooperatively with dam project organizations and with dam managers as they try to enlist the participation of downstream municipalities, residents and concerned administrative organizations.

The preparation of Reservoir Area Visions began in 2001 and by the end of 2004, visions had been enacted for more than 70% of the 99 dams targeted. This system is expected to gradually yield positive results.

The Water Resources Environment Technology Center began conducting Dam Reservoir Area Support Projects [117] in 2000 to assist groups with on-going activation activities in reservoir areas. The Center also publishes the Reservoir Area Net: a periodical that links reservoir areas with beneficiary cities.

3.3 IMPACTS ON THE RIVER ENVIRONMENT AND MEASURES AGAINST THESE IMPACTS

A dam functions by discharging the water which has stored in its reservoir and controlling the river flow rate downstream from the dam, thus forming reservoirs on rivers, resulting in artificial changes to the physical environment: flow regime, sedimentation and the water temperature and quality.

This section introduces concrete examples of measures taken by dam managers: the adjustment of water usage in response to a change in the flow regime and other measures to deal with the impact of sedimentation, water temperature, water quality, etc. on the environment.

3.3.1 Impacts of a change in the flow regime and measures against these impacts

3.3.1.1 Characteristics of river water usage in Japan [118]

In Japan, by the middle Edo Period (1603–1867), in regions where agriculture developed, the entire droughty water discharge was taken and used as irrigation water. However, during periods when rice production yields supported the region's economic strength and its population, the residents eagerly developed new paddy fields during successive years when the flow rate was high. As a result, during dry years, conflict surrounding the intake of water occurred between farmers of paddy and dry fields developed in different years, between people in upper and lower reaches, and between the left and right riverbanks in the same water supply region. Consequently, points of agreement accumulated, forming the concept of customary water use. When the quantity of water taken in and the intake periods were changed, or when new parties desiring to use the water appeared, it was necessary to form a new water usage order, and water users themselves usually performed the necessary adjustments.

The reason is one characteristic of irrigation water use, namely the fact that as a result of awareness that for each paddy and dry field, water use is closely linked to water usage systems and to private assets on each paddy and dry field, public intrusion is not possible.

In the first half of the 20th century, when the construction of dams for waterworks and hydropower began in upstream parts of rivers in regions where agricultural production was carried on with small quantities of river water in the downstream area, conflicts over water use adjustment occurred between dam organizations and downstream users.

3.3.1.2 Impacts of hydropower dams on downstream river water usage and measures against these impacts

In the case of hydropower dams, because these were initially run-of-river type, after generating power, the water was quickly discharged downstream without any reduction in quantity, so no conflict was generated between the electric power company and the downstream irrigation water users. However, when the construction of large capacity reservoir-type hydropower dams began in the mid 1920s, serious confrontations and conflicts occurred between operators of hydropower dams, who adjusted the flow rate in order to perform seasonal or daily adjustments, and downstream irrigation water users who need to consume water at a constant rate throughout the day.

One instance of adjustment where a request for compensation was made to an upstream hydropower project in relation to intake unification[3] of irrigation water, is the intake unification of the Nimangoku Irrigation System and 13 other irrigation systems irrigating 12,000 hectares on the Shogawa Fan by the Sho River Unified Intake Weir (Figure 3.3.1), which was constructed and linked to the construction of the Komaki Dam (PG, 79.2 m, #59) in the Sho River upstream.

Work on irrigation intake unification on the Sho River began in 1936 and was completed in 1940. An irrigation/drainage trunk channel-improvement project was included in this project, and compensation for the construction of the upstream hydropower plant was applied to cover local costs [119].

Prior to this, serious problems relating to water quantity occurred between the Oi Dam on the Kiso River and downstream water users of the Kottsu Irrigation System (5,000 ha) and Miyata Irrigation Systems (12,000 ha) and other users who irrigated a plain with an area of about 20,000 hectares.

These agricultural irrigation systems applied intake methods that were dependent on the stability of river flow rates based on abundant natural flow instead of intake weirs, so after the completion of the Oi Dam, the flow rate fluctuated significantly, making intake difficult and driving maintenance costs up: dredging in front of the Miyata Irrigation System intakes, repairing sluice ways and so on. A portion of such costs was covered by the electric power company that managed the dam.

In response, the electric power company took radical measures. In 1939, it constructed the Imawatari Dam (PG, 34.3 m, #72) at the confluence of the Kiso River and the Hida River, restoring the quantity discharged at the two power plants directly

---

3 Intake unification: On a river with many adjacent intakes, a unified intake facility is created by closing or combining many inlets. This is done to stabilize the intake, rationalize water usage, and reduce maintenance costs, etc. at locations with suitable conditions.

Presented by KEPCO

Figure 3.3.1 Sho River Unified Intake Weir. (See colour plate section)

upstream (Kiso River: Kaneyama Hydropower Station, Hida River: Kawabe Hydro-power Station) to a flow rate almost identical to the natural flow rate, which performed re-regulating operations to discharge the water uniformly.

Flow rate fluctuation problems occurred, not only on the Kiso River, but also on many other rivers where dams were constructed. New re-regulating ponds were constructed at only a few dams to overcome this problem, but the problem was often solved by introducing re-regulation operations at dams located downstream [120].

As one example, the Myoken Weir (#53) on the Shinano River fulfills the role of a re-regulating pond that stores water when the flow rate is high and conversely discharges stored water when the flow rate is low, in order to ensure a uniform flow of the water discharged by power production, which is done to supply electricity to commuter trains in the morning and evening rush hour at the JR Shinano River Hydropower Station located upstream, as shown by Figure 3.3.2.

3.3.1.3 The impact of dams for municipal water on downstream river water usage and measures against this impact

Since the Meiji Period, the rising water demand for waterworks in urban regions has been accompanied by increasing confrontations over the intake of the stable flow of river water between waterworks operators who wish to take in water upstream and agricultural water users downstream. The quantity of water supplied to waterworks is relatively small, excluding that in Tokyo and other large metropolitan regions, so generally adjustments are done by digging wells or constructing small reservoirs.

In waterworks systems during that period, water was often conveyed from reservoir areas directly to purification plants by pipelines to protect water quality, and this was also a method of preventing the intake of agricultural water during reservoir discharge into rivers during droughts.

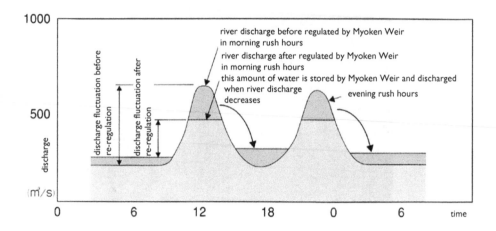

Figure 3.3.2 Re-regulating river discharge method from the JR Power Plant.

Source: MLIT [121].

In Tokyo, the quantity taken in also escalated and the construction of dams for waterworks was accompanied by serious problems relating to adjustment with downstream water users.

Figure 3.3.3 shows a water intake system on the Tama River System. Waterworks supply systems in Tokyo have been dependent on water taken in by the Hamura Weir on the Tama River since the Edo Period, but the Murayama Upper Reservoir, the Murayama Lower Reservoir and the Yamaguchi Reservoir were constructed in 1924, 1928, and 1933 respectively, thereby adjusting demand and supply with regulating ponds.

The Ogochi Dam was already planned at that time in response to further rise in water demand, but because this plan increased intake at the Hamura Weir by 5 m³/s, a serious water usage adjustment problem occurred between agricultural users of the Nikaryo Irrigation System, which takes in water on the right bank at the furthest downstream point to irrigate approximately 2,000 ha, and waterworks operators that take in water further upstream. In 1937, a compromise agreement was reached: during the irrigation period, the waterworks operator released a maintenance flow of 2 m³/s from the Hamura Weir and at the same time paid for the cost of improvements of the Nikaryo Irrigation System facilities [122].

Wide area conveyance of water during the postwar period of rapid economic growth required larger scale water usage adjustments. Tokyo planned to take in water from the upstream Tone River in order to deal with rising water demand, after 1955. However, because the Tone River supplied water used to irrigate paddy field zones in prefectures along its banks, it was necessary for agricultural water users to adjust their use of water in order to ensure a stable intake of agricultural water premised on the assurance of the flow rate required downstream, by providing a supplementary supply from upstream dams.

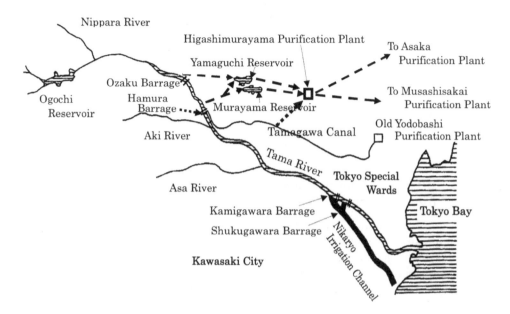

Figure 3.3.3 Schematic diagram of the Tama River and its water use.

Note: revision made on the original figure (K. Hanayama, T. Fuse, 1977 [122]).

To perform this water usage adjustment, the Gunma Irrigation System was newly constructed, and in the excessively acidic Agatsuma River, neutralization facilities were constructed on the upper reach of the Tone River to improve the water quality. Downstream, the Tone River Estuary Weir was constructed and was capable of stable intake without significantly being affected by the salt concentration in the Tone River, even during droughts. An unified intake weir (the Tone Barrage) was constructed in the middle reach of the river, to take in water for both new urban use and agricultural use, thereby resolving the water usage adjustment problem by about 1962 [123].

3.3.1.4 Maintenance of the normal functions of river flow by river managers

The revised River Law of 1964 stipulated the flow rate necessary to maintain the normal functions of river flow at key points. At the Fujiwara Dam and the Aimata Dam, which had already been constructed on the Tone River, the capacity was stipulated as the "capacity for existing irrigation".

This flow rate needed to maintain the normal functions of river flow is called the "normal flow rate[4]", and consists of the water use flow rate that is necessary for downstream water use and the maintenance flow rate that should be maintained during a dry season. The normal flow rate not only ensures the water use flow rate, it also plays major roles in conserving and restoring habitats and breeding grounds for fauna and flora, maintaining and improving river scenery, preserving the cleanliness of flowing water, transporting boats, preventing salt damage, preventing river mouth blockage, protecting river management structures, maintaining the groundwater level and ensuring the existence of recreational land [124, 125].

In the upstream section of the main course of the Tone River, in 2000, a capacity of 150 million m^3 of water for normal river function was ensured during the winter by the above-mentioned group of dams.

Because of the past priority use of water for human activities such as irrigation, producing electric power and supplying waterworks systems on rivers with an insufficient maintenance flow rate to provide habitats for fauna and flora, studies should be done to ensure a normal flow rate.

3.3.1.5   Discharging the river maintenance flow rate from hydropower dams

In the case of a dam and conduit type hydropower plant, river water is taken in at the dam intake on the upstream river and is then conveyed to the hydropower plant where it is discharged downstream, thus drying up the river between the reservoir and the discharge outlet point (dried-up river), which destroys the river's original environment. In the 1980s, strong demands from local people for beautiful rivers filled with abundant clear water began, and the number of instances in which it was difficult to meet these demands increased.

The MLIT, which is the ministry responsible for river management, has, through consultations and coordination with the METI that has jurisdiction over hydropower production, responded to this problem by announcing a notification ensuring the river maintenance flow rate in 1988. This notification stipulates that when each electric power plant renews its water rights for electric power generation use, it must begin to discharge a stipulated maintenance flow into the downstream river from the hydropower dam.

As a result, by the end of March 2002, in line with the notification, clean flowing water had been restored in dried-up river sections at 287 electric power plants, and as shown in Figure 3.3.4, the total section length was 3,500 km, indicating that about 37% of a total of 9,600 km of dried-up river courses had been improved [126]. Furthermore, continued improvements are anticipated. By March 2002, the discharged flow rate reached an average of about 0.3 m^3/s per 100 km^2 in each river's catchment area.

3.3.2   The progress of sedimentation in dam reservoirs and measures against this problem

3.3.2.1   The state of sedimentation in reservoirs

The volume of sediment at the end of 2001 in 911 reservoirs in Japan was, as shown in Figure 3.3.5, an average of only about 7% of total reservoir capacity, but there

---

4 Normal flow rate: The flow rate necessary to maintain the normal functions of river flow. This flow rate satisfies both the maintenance flow rate, stipulated as that which should be maintained during a dry season, and the water use flow rate necessary for use of flowing water in the lower reach of a river.

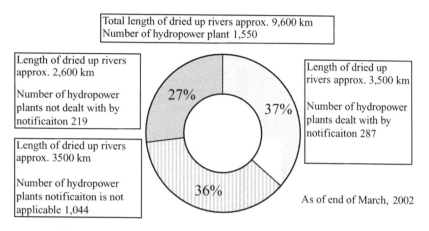

Figure 3.3.4 **Length of river courses restored from dried up condition under notification (Class A River Systems).**

Source: MLIT [126].

were great differences according to the region, river system, etc. On rivers flowing near a tectonic line or a fault dividing a geological structure, many reservoirs suffer from advanced sedimentation because of the large quantities of sediment that flow into them. The quantity of sediment by region shows that it is heaviest in Chubu Region, followed by Hokuriku Region, with sediment amount in both regions accounting for 52% of the national total.

Among hydropower reservoirs that were either completed or planned prior to 1957, over 40 had sedimentation rates of more than 50%, and some were nearly full.

3.3.2.2 Impacts of reservoir sedimentation

Dams are constructed to store water, but reservoirs also store the sediment contained in water.

The advance of sedimentation of a reservoir directly impacts the dam's functionality by reducing reservoir capacity and obstructing intake and at the same time the sediment deposited near the upstream end of the reservoir raises the riverbed. Additionally, a dam's disruption of the continuity of the movement of sediment lowers the downstream riverbed and erodes the coastline.

Sedimentation of the Yasuoka Dam on the Tenryu River that has had impacts extending to the riverbed and flood discharge level in the region about 10 km upstream from the dam, began causing problems around 1940. There is a narrow portion of river located between these regions and the Yasuoka Dam downstream (a famous scenic spot called Tenryu-kyo Gorge). During a flood, the water level rises at this narrow spot, but this topography is not the only cause: it is impossible to ignore the impact of the rapid sedimentation of the Yasuoka Dam (see Figure 3.3.6) on flood disasters that have occurred since about 1940.

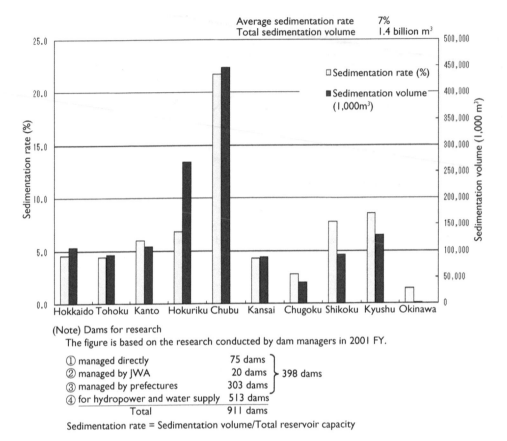

(Note) Dams for research
The figure is based on the research conducted by dam managers in 2001 FY.

① managed directly	75 dams	⎫
② managed by JWA	20 dams	⎬ 398 dams
③ managed by prefectures	303 dams	⎭
④ for hydropower and water supply	513 dams	
Total	911 dams	

Sedimentation rate = Sedimentation volume/Total reservoir capacity

Figure 3.3.5 Sediment volume and rate by region (2001).

Source: MLIT [127].

After the construction of the Yasuoka Dam, floods caused severe damage in 1961 and 1983. The 1961 flood disaster, in particular, accelerated the trend towards demanding full-scale flood control measures.

In response to this trend, at the Yasuoka Dam, gravel sediment has been excavated near the upstream end of the reservoir since 1984. At the same time, a project was carried out to prevent flood damage by performing landfill in a region inundated by a flood about 10 km upstream from the dam, thereby raising the ground level by about 6 m [128].

3.3.2.3   Sedimentation countermeasures [129]

In Japan, planned sedimentation capacities have been set since the 1950s to prepare for the decline in reservoir functionality by ensuring an approximately 100-year reservoir sedimentation capacity in advance as the planned sedimentation capacity within

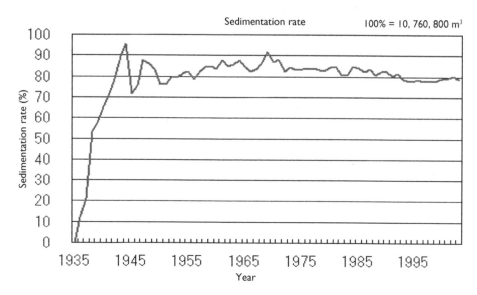

Figure 3.3.6 Change of sedimentation rate at Yasuoka reservoir.

Source: CEPCO.

a reservoir. Since the late 1960s, reservoirs have been periodically surveyed to measure sedimentation within the reservoir capacity.

Sedimentation measures are broadly categorized into the following three types. Examples of these countermeasures are introduced, but many problems that will require future technological development remain, such as cost problems and river environment problems.

3.3.2.3.1   Measures based on excavation or dredging and removal

The most widely applied sedimentation countermeasure adopted at many dams is excavation or dredging, followed by removal of the sediment material. In many instances, a check dam is constructed and sediment collected by it is excavated and removed.

According to a 1998 questionnaire survey of dam managers across Japan by the MLIT [130], of the 580 dam managers who responded, about 25% indicated that sediment in reservoirs is excavated and removed. During the decade from 1988 to 1997, an average of about 3.9 million m³/year of sediment was removed from reservoirs, and the most removed in one year was 5 million m³, mostly excavated by backhoes and dredging barges and then transported by dump truck, in many cases for distances of up to about 20 km.

At 114 (78%) of 147 dams, the sediment removed was used effectively, and of the approximate 3.9 million m³ removed, approximately 2.2 million m³ was used effectively. About half was used as concrete aggregate, but in recent years, its use as soil improvement material or as material returned to rivers to mitigate riverbed degradation has attracted attention.

A typical example of sedimentation countermeasures taken, based on excavation or dredging and removal, are those taken at the Sakuma Dam Reservoir.

Since about 1970, it has been necessary to protect the surrounding region from inundation caused by the rise in the riverbed in the sedimentation area upstream from the reservoir during floods. The following sedimentation countermeasures are now taken [129, 131].

- Transportation outside the reservoir area: Since about 1970, private companies have dredged the reservoir to obtain construction materials.
- Transportation within the reservoir area: Since 1990, the dam manager, J-POWER, has dredged the reservoir, moving the material into the dead water capacity (deeper than the effective capacity) inside the reservoir instead of transporting it outside the reservoir area.
- Acceleration of sand flushing: Since 1991, the dam manager has lowered the reservoir water level during the dry season and used the flowing water during small river flow rate periods to move the sediment into the dead water capacity.

In addition, as part of a comprehensive river system management, the 1998 report by the Subcommittee of Sedimentation Management of River Council clearly showed the need for comprehensive sedimentation management based on the "sediment transportation system" concept. Since then it has become necessary for the problem of sedimentation of reservoirs to be included in such a comprehensive system.

Under such background circumstances, as part of comprehensive sediment management, efforts are being made to restore sediment in the downstream river below dams such as the Akiha Dam, Miharu Dam (PG, 65.0 m, #21), Futase Dam, and the Shimokubo Dam [132]. Part of the sediment deposited in the reservoir is excavated then placed inside the river downstream from the dam so that it is transported by flowing water during flood periods. Removing sediment which would probably have been carried downstream by the flowing water without the dam, then restoring it to the downstream river can be counted on to restrict riverbed degradation in the downstream river, prevent the coarsening of the downstream riverbed material, and improve habitat environments for fish.

3.3.2.3.2   A method based on a sand flushing system in dam bodies

The Kurobe River has a higher specific sediment yield rate than any other river in Japan. Therefore, at the Dashidaira Dam (PG, 76.7 m, #55, Figure 3.3.7) and the Unazuki Dam, large sand flushing conduits and sand flushing gates have been installed in the dam bodies and the reservoir water level is lowered to increase tractive force, as a means of flushing deposited sediment from inside the reservoir.

This facility was used for the first time in 1991, six years after the completion of the Dashidaira Dam. The flushed sediment which contained organic material produced by anaerobic decomposition at the bottom of the reservoir caused deterioration in the water quality in the downstream region of the river. Later, the flushing method was

Presented by KEPCO

Figure 3.3.7 Sand flushing at the Dashidaira Dam. (See colour plate section)

improved, almost achieving a method of flushing sediment in a way similar to natural flooding. After completion of the Unazuki Dam about 7 km downstream from the Dashidaira Dam, a committee was established to carry out coordinated sediment flushing and sluicing by the two dams, beginning in 2001.

3.3.2.3.3   A method based on a sand bypass tunnel

In the case of the Asahi Dam (VA, 86.1 m, #100), sand bypass started in 1998. This is discussed later because this was mainly intended to be a turbidity prolongation countermeasure.

The MLIT started the Miwa Dam (PG, 69.1 m, #65) Redevelopment Project in 1989. It excavated sediment already deposited inside the effective capacity (4.2 million m^3) and at the same time constructed permanent sedimentation countermeasure systems including check dams, diversion weirs, and sand bypass tunnels, etc. The sand bypass tunnel has a total length of about 4.3 km and maximum capacity of 300 m^3/s (section area 47 m^2), and after the check dam has captured coarse sediment, it diverts flood discharge, including wash load, to convey it directly below the dam detouring the reservoir. Trial operation of this facility was done in June 2005.

3.3.2.4   *Ensuring sound continuity with material circulation*

The impacts of dams on river environments have been assumed to be related to the flow regime and sedimentation, and measures to mitigate these effects are taken. Countermeasures are also taken to mitigate impacts on water quality, water temperature and living organisms, and these are discussed on the next paragraph and in the next section. Measures to deal with elements such as water, sediment and living organisms have been taken, and in recent years in particular, there has been a growing awareness that rivers play an important role as transport routes in material circulation between inland areas and the ocean. Dams that form reservoirs actually change the continuity of rivers between the mountains and the ocean, and it is necessary to clarify the impacts of dams on the movement and circulation of nutrient salts and other materials [133].

At this time, little is known about material circulation, but ensuring the sound continuity of materials in a river system is an important challenge.

3.3.3   Measures against water temperature and water quality problems in reservoirs

3.3.3.1   Cold water problems and measures against these problems

When water is taken from the middle and deeper layers of a reservoir, the water with lower temperature is discharged. At conduit type hydropower plants, water is conveyed through long conduits, so cold water is discharged into the river without its temperature rising as it does in a natural river flow. As a result, it is either impossible to supply irrigation water of a temperature suitable for growing paddy rice, or it adversely affects downstream fisheries. Since the mid 1920s, resolving the problem of cold water caused by hydropower plants in this way has been a challenge.

Since the mid 1940s, dam projects have often harmed rice production by releasing cold water in mountainous areas and upper latitudes [134].

Measures based on equipment inside reservoirs have been taken since approximately 1953: providing surface intake systems that can selectively take in warm water near the surface by installing intake gates that follow the water level and multiple inlets.

The Ikawa Dam constructed on the Oi River is introduced as an example of their use. Under the Oi River Agricultural Water Use Plan enacted by the MAFF and the plan for Kawaguchi Hydropower Station, CEPCO, based on the Sasamagawa Dam (PG, 46.4 m, #81), water discharged by the Ikawa Dam flows down an approximately 43 km long tunnel that conveys it directly into the irrigation channel constructed by the Oi River Agricultural Water Use Project. As a result, there was a danger that the rice harvest might decline because the temperature of irrigation water was lower than that of water taken in from the main course of the Oi River.

Based on studies, it was concluded that the water temperature would be restored by moving the intake ports at the Ikawa Dam from a bottom to a surface intake mode at a depth of 10 m, so this method was adopted.

In the 1970s, a selective intake[5] was introduced as a countermeasure against problems such as turbidity and eutrophication, as described below.

At the Sagae Dam (ER, 112.0 m, #19), two selective intake gates were installed to take in the maximum intake of 62.5 m³/s. Basically, warm water is taken from the surface layer, but as a measure to discharge warm water over a long period and to discharge turbid water after a flood, water can be selectively taken in from two layers – the surface and middle layers – and can be mixed, if necessary.

Recent measures taken to prevent the discharge of cold water from dams have often been implemented using selective (surface) intakes.

3.3.3.2 The problem of prolonged turbidity and measures against the problem [135]

Extremely fine particles that have flowed into a reservoir during flood discharge are occasionally suspended inside the reservoir for a long time. When a reservoir becomes turbid in this way, taking water from it and discharging water downstream damages not only the downstream habitat environments of living organisms and landscapes, but also the downstream sweetfish fishery and the water supply intake and water-works facilities, etc.

As an example of prolonged turbidity problems in Japan, the Sameura Dam (which began operating in 1973) can be cited as an example of a case where this phenomenon was caused by a flood triggered by a typhoon in the fall of 1976.

The discharge turbidity was still more than 100 degrees in mid-October more than a month since the typhoon, and it did not fall to 10 degrees until late November.

Such a prolonged turbidity phenomenon in reservoirs is, as a hydraulic phenomenon in a reservoir, explained as follows.

When a large flood discharge flows into a reservoir, all the water in the reservoir becomes turbid, prolonging the discharge of turbid water after the flood. Turbid materials that have flowed in during the summer flood discharge gradually settle to lower layers. Then in the autumn and winter, the chilling of the surface water encourages vertical circulation in the reservoir, resulting in transportation of turbid bottom water to the surface layer. Thus even in winter, six months after the flood, this circulation phenomenon continues causing high turbidity [136].

Measures to prevent prolonged turbidity are broadly categorized as production source measures, in-river measures, and in-reservoir measures, according to the process: The production, transport and storage of turbid materials.

Examples of these measures are introduced below.

3.3.3.2.1 Production source measures

Production source measures, that reduce turbidity by restricting the quantity of turbid materials produced, include the following types.

- Developing forests to diversify tree species and forest physiognomies.
- Sabo works and afforestation that prevents landslides and the movement of unstable soil.

---

5 Selective intake: An intake that can take in water from either the surface, middle, or deep layers of a reservoir. To deal with water quality problems in reservoirs such as discharge of cold or turbid water and eutrophication, selective intake is connected to the water use outlet to perform intake operations that use the density stratification of the reservoir.

3.3.3.2.2   In-river measures [137]

In-river measures prevent the inflow of turbid materials from the drainage basin into reservoirs. A major method in this group is a detour channel: when turbidity is high in a downstream river during a flood, turbid water flowing from upstream towards the reservoir is transported directly downstream without entering the reservoir, thus preventing turbid materials from flowing into the reservoir. After the flood, transporting clear water from upstream into the river downstream from the dam through this channel, reduces turbidity in the downstream river area.

At the Asahi Dam, a sand bypass tunnel started operating in 1998. The tunnel is about 2.4 km long and its maximum capacity is about 140 m³/s (a section area of 10 m²). And it reduces prolonged turbidity in the reservoir by transporting turbid water from the upstream end of the reservoir through the bypass tunnel directly into the river downstream from the Asahi Dam during a flood, thereby contributing to mitigating turbid water problems in the upstream part of the Shingu River System (Figure 3.3.8).

Like the reservoir detour channel at the Kamiosu Dam (ER, 98.0 m, #74) of the Okumino Pumped Storage Power Plant, some systems function comprehensively as turbid water countermeasures, maintenance flow rate discharge systems, and as in-house power generating equipment of the Okumino Power Plant.

3.3.3.2.3   In-reservoir countermeasures

In-reservoir countermeasures include the following:

- Selective intake to lower turbid discharge by discharge operations.
- Channel work at the upstream end of reservoirs to prevent scouring of deposited sediment when the reservoir water level falls.
- Dredging and excavation of sediment deposited at the upstream end of the reservoir.

Selective intakes have been installed at many dams in recent years. In this section, examples of measures taken at the Sameura Dam are introduced, where, as stated above, turbidity prolongation phenomena have caused social problems.

At the Sameura Dam, the surface intake method of taking in clear water at the surface had been adopted, but the installation of selective intakes was proposed and the existing surface intake was reconstructed in 1999, to build a selective intake that can freely select the location from which the water is taken in.

This system provided a maximum intake of 70 m³/s, and as shown in Figure 3.3.9, its variability range is 38.20 m. When turbidity is high near the reservoir surface during a flood, making it impossible to take in clear water from the surface, selective intake can be performed.

In this way, at the Sameura Dam, the selective intake is used to reduce turbidity by dealing with it by the early discharge of turbid influent.

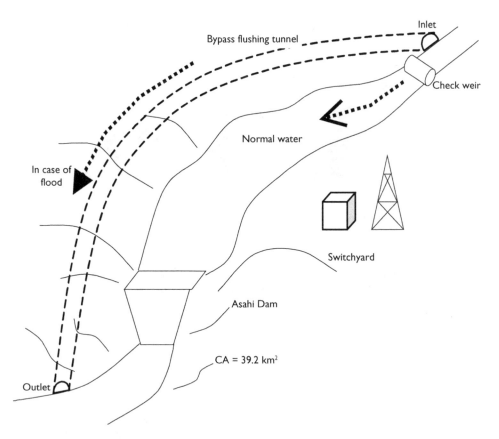

Figure 3.3.8 Outline of the sand bypass tunnel.

Source: Japan society of Dam Engineers [135].

3.3.3.3 *Eutrophication phenomenon in reservoirs and measures against this phenomenon*

Inside reservoirs, water is stored for a certain period and there are, therefore, instances where the eutrophication phenomenon caused by abnormal propagation of algae occurs according to the concentration of nitrogen, phosphorus and other nutrient salts flowing from the dam catchment area, the rotation rate of reservoirs, and weather conditions in the vicinity of the reservoir, etc. When eutrophication has occurred, discharge from the dam may harm the quality of water in the downstream river, causing problems such as unpleasant odors in the water supply and clogging or other problems in filtration ponds at purification plants.

One instance where eutrophication in a reservoir in Japan has caused serious social problems is the case at the Kamafusa Dam (PG, 45.5 m, #17), which is the water source for the waterworks in Sendai City. At the Kamafusa Dam, malodorous water

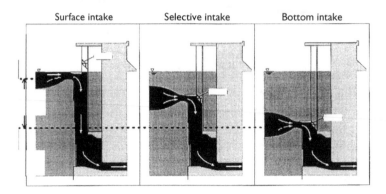

| Surface intake | Selective intake | Bottom intake |

Figure 3.3.9 Outline of intake methods.

Source: MLIT.

(with a musty smell) occurred frequently in 8 of the 13 years, from 1970 when first filling started to 1983. The Sendai City Waterworks Bureau tried using powdered activated carbon treatment to cope with this problem. The treatment was performed on an average of 130 days/year at an annual cost of 60 million yen, placing a heavy financial burden on the city [138].

Among measures taken to prevent the eutrophication phenomenon, aeration circulation devices[6] and partition fences, etc. are installed as in-reservoir measures, and are combined with fore-pond or other facilities that process inflowing water. To reduce the flow of nutrient salts into a reservoir, support is provided to encourage the construction of sewage systems as anti-pollution source measures in the river basin upstream from the reservoir, and measures are taken to discharge nutrient salts that flow into a reservoir from the river directly downstream from the dam.

Aeration circulation measures, which restrict the propagation of phytoplankton inside reservoirs, are taken at many reservoirs. The ascending power of air bubbles to vertically circulate reservoir water is used to restrict phytoplankton propagation by controlling light and water temperature, and to provide the nutrient salt supply restriction effects of flow control [139, 140].

At the Kamafusa Dam, the total layer circulation method using an aerated circulation system has been applied since 1984. In 1987, it became the first reservoir where a lake water quality conservation plan was prepared based on a designation under the Law Concerning Special Measures for Conservation of Lake Water Quality. As a result of these measures, the COD inside the reservoir fell, controlling musty odors [141], but in 1996, this musty odor recurred and from 2003, diffusive aeration[7] circulation devices were introduced in stages to conserve water quality.

---

6 Aeration circulation device: A device that ejects air inside a reservoir to cause a circulating flow in the vertical direction. Types of aerated circulation are shallow layer aeration, which is aerated circulation in a shallow layer to disrupt surface stratification, and deep layer aeration, which causes deep layer water to circulate without disrupting stratification.

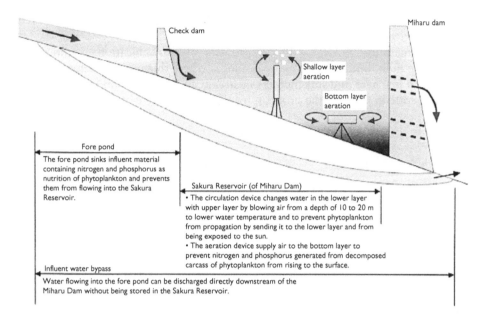

Figure 3.3.10 Outline of water quality conservation measures at the Miharu Dam.

Source: MLIT [144].

At the Miharu Dam (PG, H = 65.0 m, MLIT, 1997), as shown in Figure 3.3.8, a wide range of eutrophication countermeasures have been taken [142, 143].

- Influent water bypass and fore-ponds (4 locations) to restrict the inflow of nutrient salts to the reservoir.
- Shallow layer aeration (5 locations) to restrict propagation of algae.
- Bottom layer aeration (2 locations) to prevent the production of an anaerobic water mass in the bottom layer.

There are aspects of the effectiveness of aeration circulation that are difficult to quantify, so further studies must be performed of their countermeasure scale, effectiveness and efficient methods of operating them.

Other measures that are being used increasingly in recent years along with aeration circulation include the partition fence method: dividing the upstream and downstream parts of reservoirs with lateral membranes at a depth from 5 m to 15 m to control the flow of the reservoir's surface layer (Figure 3.3.11, Figure 3.3.12). This has been introduced at many dams: the Hachisu Dam (PG, 78.0 m, #87), Shimajigawa Dam (PG, 89.0 m, #105), Ishidegawa Dam (PG, 87.0 m, #114), Shimouke Dam, Terauchi Dam (ER, 83.0 m, #119), Hinachi Dam, Hitokura Dam, and the Urayama Dam [145–147].

---

7 Diffusive aeration: Ejecting compressed air as fine bubbles from a blower, to cause a circulating flow and supply oxygen.

Other eutrophication countermeasures are soil treatment and plant purification, using aquatic plants or periphyton and influent water bypass systems, etc. to reduce influent nutrient salts [148, 149].

To take measures to deal with the problem of eutrophication, countermeasure facilities are built, but there are many instances where this problem remains.

Each dam manager and stakeholders upstream and downstream from the reservoir must cooperate in tackling the problem from the perspective of entire river basins.

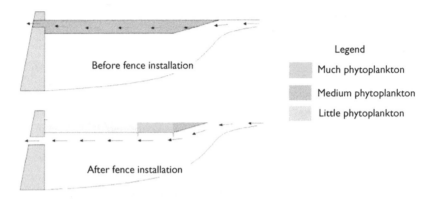

Figure 3.3.11 Function of reservoir partition fence (with and without). (See colour plate section)
Source: JWA.

Figure 3.3.12 Installation of reservoir partition fence. (See colour plate section)

3.4 IMPACTS ON HABITATS OF LIVING ORGANISMS AND MEASURES AGAINST THESE IMPACTS

Constructing a dam brings sweeping changes to the topography, not only the creation of a reservoir, but also the construction of diversion roads, etc. These changes affect the natural environment in a variety of ways.

In recent years, concern regarding the conservation of the natural environment as a way to realize ideal biodiversity and recycling-based sustainable development has grown. To conserve the habitats of living organisms at dam project sites, various actions have been taken to protect valuable species and conserve the natural environment around reservoir areas, flexible dam management methods have been tested, and so on.

3.4.1 Impacts on the habitats of living organisms by dam projects

3.4.1.1 Expanding awareness of impacts on ecosystems

In 1988, when full-scale work began on an estuary weir on the Nagara River (the Nagaragawa Estuary Barrage) in Chubu Region, the League to Oppose the Nagaragawa Estuary Barrage was formed and commenced its campaign called, "Protect the one remaining natural river, the Nagara River! It will wipe out the Satsukimasu cherry salmon!" In response to this appeal, the L Nature Conservation Society of Japan, the Wild Bird Society of Japan, the Japanese Committee of the WWF, and other nature protection bodies and concerned academic groups also voiced their objections one after another.

The project's organizers (MLIT and JWA) responded by conducting a series of broader environmental impact surveys and disclosing all findings to clarify the facts.

The Nagaragawa Estuary Barrgae issue reveals growing public concern about the impact of dams on the natural environment.

These circumstances were reminders of the various ways in which dam construction affects the natural environment, because the formation of the dam reservoir, excavation of the quarry, construction of diversion roads and so on, largely transform the topography. Specifically, the formation of a dam reservoir reduces or destroys the habitats and breeding environments of fauna and flora while the dam breaks the upstream – downstream continuity of the river, dividing habitats and breeding environments of fauna and flora.

In recent years, awareness has grown (see Figure 3.4.1) that ecosystems in the natural world are the foundation of the growth of plants, algae and phytoplankton caused by nitrogen, phosphorus and other nutrient salts supplied from upper reaches of rivers and by solar energy, and that these form hierarchical food chains. In this way, the importance of material circulation in river environments has been widely recognized.

3.4.1.2 Impacts on fisheries

In 1960, a periodical that is read by members of the general public and that reports on sweetfish[8], which is a typical freshwater fish in Japan, published articles containing

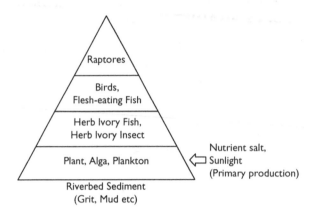

Figure 3.4.1 Ecosystem pyramids – the water/material relationship.

Source: S. Ikeda, Engineering for Dams, 2001 [150].

the following statements: "When a stream is changed into an artificial lake by a dam intended to generate electric power or to control floods, sweetfish can no longer inhabit parts of the reservoir, so the fishing industry must be compensated." The report continued, "In the midstream part of a river, far more sweetfish are produced than any other kind of fish. Therefore, when calculating compensation, it is always important to find out how many sweetfish are being caught" [151]. This reveals the views prevailing at that time on the impacts of dam construction on fisheries.

The Lake Biwa Biotic Resources Investigation Group (1961–) formed before the Lake Biwa Comprehensive Development Project and the Investigation of Estuary Resources of the Kiso-Nagara-Ibi Rivers (1963–) [152] carried out before the finalization of the Nagaragawa Estuary Barrage Plan have provided much valuable information about the ecosystem of the sweetfish, from the perspective of resource surveys.

3.4.2   Social trends concerning natural environment conservation and dam projects

3.4.2.1   *Change in environmental problems [153, 154]*

During the postwar rehabilitation and rapid economic growth periods, severe public nuisance problems, including air pollution, water pollution and ground subsidence appeared. These were accompanied by an interest in preventing pollution caused by

---

8 Sweetfish (*Plecoglossus altivelis*): A fish that migrates between rivers and the ocean, which is typically caught in narrow mountain rivers and is an important food source in Japan. Famous methods of catching sweetfish include using tamed cormorants (Pelicaniformes), and *tomozuri* (this fishing method involves attaching a live sweetfish to a fishing line, attaching a hook to its stomach, then lowering it into the river. When a free-swimming sweetfish tries to protect its feeding territory by attacking the stomach of the sweetfish being used as a decoy, it becomes caught on the hook attached to the decoy's stomach. This method takes advantage of the sweetfish's territorial instinct).

construction work at dam projects. The government therefore enacted the Basic Law for Environmental Pollution Control in 1967, the Water Pollution Control Law in 1970, and the Nature Conservation Law in 1972, thus establishing a system of laws concerning pollution and environmental conservation measures centered on regulatory measures.

During the 1970s, two oil shocks ended Japan's rapid economic growth period and ushered in a period of economic maturity, thereby stimulating people's environmental awareness as they enjoyed more free time and their values diversified. People's interest in improving the environment surrounding reservoir areas and using dam reservoirs also developed.

Since the late 1980s, the public has become even more aware of the seriousness of global warming, the destruction of the ozone layer and other global environmental problems. The spread of mass production, consumption and disposal lifestyles have created new environmental problems, such as waste and recycling problems. Endocrine disruptors (environmental hormones) and rare species recorded in the Red Data Book[9] also attract interest. In addition to pollution prevention, protection of the environment from their impacts was added to the list of goals for dam projects as the public became increasingly sensitive to environmental conservation.

In 1992, the United Nations Conference on Environment and Development (UNCED) (Earth Summit) was held in Rio de Janeiro in Brazil, drawing the world's attention to global environmental problems. The Earth Summit attempted to establish concrete countermeasures to conserve the global environment and achieve sustainable development – both are common problems of the world's people – and called on participating countries to respond to threats to the global environment.

Japan's national consciousness of the environment was also largely transformed. For example, according to the Public Opinion Poll on the Protection and Utilization of Nature carried out in 2001, 40.1% of respondents answered, "Conserving nature is most important so that people can live" and the percentage giving this answer has risen steadily (it was 28.5% in a 1986 survey). Under these circumstances, in recent years, interest has focused on conservation of the natural environment under the concept of biodiversity and change to the recycle-based sustainable economic society.

3.4.2.2 Environmental impact assessments

Environmental impact assessments (EIA) have been carried out by organizations implementing individual projects since 1972. However, procedures were not unified and assessment methods were not fully established, so since 1984, EIA based on unified rules have been carried out at large-scale projects, with the participation of the central government. In 1999, the Environmental Impact Assessment Law was enforced, thereby creating procedures necessary to respond to the demand for conservation of the environment that appeared at that time.

---

9 Red Data Book: A list of living species that should be protected because they are considered to be in danger of extinction by the International Union for Conservation of Nature and Natural Resources (IUCN). Japan has also prepared a Japanese version of the Red Data Book, which lists species classified as extinct, endangered, vulnerable, and rare.

Environmental conservation measures that were taken at many dam projects prior to 1972 were ineffective by today's standards. In response to public demand, later dam projects have been accompanied by an evaluation of impacts on the natural environment and the introduction of improved environment – preservation measures, based on research and development activities. Environmental conservation measures can be broadly categorized as: avoidance – preventing impacts on the environment; mitigation – minimizing such impacts; and compensation – restoring lost environments. Currently, studies are carried out that conform to the law, even for many small dam projects that are not subject to these provisions.

3.4.2.3 Revision of the River Law

River management is now required not only to provide flood control and water supply functions, but also to conserve rich natural environments and play a role in creating well endowed living environments. Therefore, in 1997, the River Law was revised by the addition of improvement and conservation of the river environment to its purposes, and by revising the river planning system.

Two reasons have been given for this revision. Firstly, when the provisions of the former River Law were stipulated, it was difficult to understand the importance of conserving river ecosystems and river scenery, considerations that have been emphasized in recent years. Secondly, to further improve and conserve the river environment in response to people's needs, it was necessary to clarify that protecting the environment is a purpose of river management.

3.4.3 Examples of efforts to preserve ecosystems

3.4.3.1 Riverfront census

Since 1990, riverfront censuses have been conducted on 109 Class A River Systems throughout Japan and at dam reservoirs of multi-purpose dams managed by the MLIT and the JWA. Since 1993, these surveys have been expanded to include major rivers and dams managed by prefectures, and as natural environment surveys that are conducted regularly, continually, and comprehensively, they provide fundamental information about dam reservoirs and river environments.

3.4.3.2 Monitoring

While a dam is being constructed, it is extremely important to take preservation measures based on EIA throughout the work period and to clarify changes in circumstances after such impact assessments. After the facility has been completed, in order to manage it appropriately it is important to clarify conditions predicted at the construction stage and to reflect on the results. Therefore, at the construction stage and during the operation stage, the facility's impact on fauna and flora is clarified by monitoring, and the findings are analyzed and evaluated.

Specific methods of monitoring during the work period are environmental patrols to survey the felling of standing trees and monitoring of illegal dumping of waste on the project site. Additionally, to protect raptors, work must be scheduled considering their breeding season. There are also instances where visual monitoring is carried out to confirm breeding, accompanied by continuous moving picture observations done with CCD cameras [155].

At the operation stage, a monitoring committee of experts is formed to conduct surveys and analyses of environmental impacts and to announce their findings in compliance with the Monitoring and Evaluation System for Dam Management (see 3.5).

3.4.3.3 Measures for valuable species

3.4.3.3.1 Measures for raptors

Golden eagles, mountain hawk eagles and other raptors are positioned at the top of ecosystem food chains. In regions inhabited by raptors, the environment is diverse and well preserved.

The numbers of raptors are dwindling and some species have been designated as endangered by the Ministry of the Environment (MOE), resulting in recent strong demands for their protection.

Many dam projects are implemented in mountainous regions, so they often affect raptors' habitats which are usually mountainous regions. Current dam construction projects are nearly always accompanied by carefully conducted surveys and studies to conserve raptors' habitats. And the work includes measures to coordinate work periods and to reduce noise and vibration. In other cases, measures such as changing the routes of diversion roads, etc., are taken to protect areas that are important for breeding purposes [156].

At the Mikawasawa Dam (PG, 48.5 m, #23) [157], for example, it was confirmed that there were mountain hawk eagles inhabiting the dam project district (their nesting tree was about 700 m from the dam), and a field survey was carried out to clarify the state of the habitat. As a result, conservation measures were taken throughout the entire work period, including the use of low noise and low vibration construction machines, controlling access to the nesting area, and taking work period related conservation measures, such as adjusting the commencement of work according to their nesting behavior. It has been confirmed that because of these efforts, the raptors nested and their fledglings left the nest during the work period.

To enact conservation measures, experts and academics formed a Mountain Hawk Eagle Conservation Committee that established sets of measures for each work category that might possibly affect breeding.

3.4.3.3.2 Measures for fish [158]

Impacts of dam construction on the habitats of fish include impacts on breeding and hatching by a change in the flow regime of the river, in the constituent riverbed materials and in water temperature and quality downstream from the dam. To reduce these impacts, the discharge of water from dams is controlled by performing flexible management discussed below and by using selective intake.

Direct impacts of dam construction are the prevention of fish migration upstream and downstream from the dam site, so as a countermeasure, fishways are provided at dams where their construction is feasible.

Fishways have been constructed at intake weirs and other river structures for many years, but many are provided for the benefit of specified fish species that are important to the fishing industry. In recent years, new emphasis on diversity and the conservation of river environments has led to proposals of new types of fishways, designed for use by many different fish species.

Fishways at dams have been constructed at more than 30 locations with a head fall of up to 30 m, as shown in Figure 3.4.2. As shown in Table 3.4.1, most of these are pool types.

When a fishway is installed at a dam, the problems faced include a high head, large reservoir water level fluctuation, and fish having difficulty descending the fishway. Extensive research and development are being done to resolve these problems.

Fishways at dams must be easy for migratory fish, such as salmon, trout and sweet-fish that migrate between the ocean and rivers, to ascend and descend. Examples of special measures to help fish descend fishways include fishways that connect the upstream end of the reservoir to the river directly downstream from the dam. Future research must develop measures to induce descending fish to enter fishways in large reservoirs.

3.4.3.3.3   Measures used for the preservation of plant species

Plant species are conserved, in some cases, by taking impact avoidance measures that begin during the planning stage, such as those taken at the Tambara Dam (ER, 116.0 m, #37) where the dam height is restricted to preserve a marshland where skunk cabbage and other high plateau vegetation can grow, thereby maintaining biodiversity. Generally, the felling of trees on sites that will be affected is postponed for as long as possible so that as many seeds as possible can disperse and take root, or sometimes the affected trees are moved and planted in other areas.

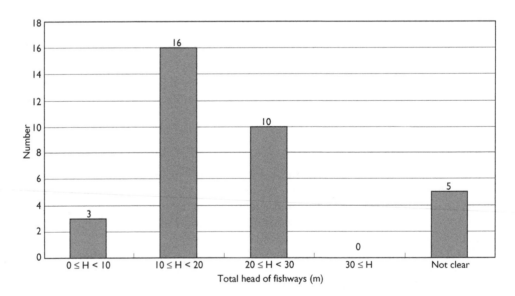

Figure 3.4.2 Distribution of total head of fishways.

Source: N. Koike, G. Saito, Report of Water Resources Environment Research Institute, 2002 [158].

Table 3.4.1 Fishways recently installed at dams in Japan.

Dam name (Completion year of fishway)	Project manager	Type	Specifications of fishways			
			Total height (m)	Slope (1/m)	Discharge (m³/s)	Distance (m)
Nibutani Dam (PG, 32.0 m, #6, Figure 3.4.3) (1997)	The Hokkaido Development Bureau	Pool and Weir Type + Sector Type	16 (31.5)	10	0.25–0.33	164
Samani Dam (PG, 44 m, #7) (1999)	Hokkaido	Pool and Weir Type	22 (44.0)	10	0.34	288
Nakaiwa Dam (VA, 26.3 m, #28) (1997)	TEPCO	IH Type + Slope with rough stone	18 (26.3)	10	0.292	157
Meboro Dam (PG, 40.0 m, #120) (2000)	Nagasaki	Pool and weir Type + Sector Type	14 (40.0)	10	0.086	153
Aono Dam (PG, 29.0 m, #93) (2001)	Hyogo	Pool and weir Type + Nature conservation Type	24 (29.0)	15–150	1–0.3	730
Shiromaru Dam (PG, 30.3 m, #48) (2001)	Tokyo	IH Type + orifice Type	27 (30.3)	10	0.28	331
Setoishi Dam (PG, 26.5 m, #132) (1999)	J-POWER	IH Type	17 (26.5)	15	0.5	432
Pirika Dam (PG/ER, 40.0 m, #5) (2005)	The Hokkaido Development Bureau	Pool and Weir Type + Nature conservation Type	30 (40.0)	15–33	0.5	2,360

Source: N. Koike, G. Saito, Report of Water Resources Environment Research Institute, 2002 [158].

At the Taiho Dam (PG, 77.5 m, #134), appropriate measures were taken, including transplanting cuttings of airy shaw (*Margaritaria indica (Dalz.)*), and the aquatic macrophyte, *Blyxa, aubertii LC Richard*, etc. on suitable growing land, where they were monitored for rooting and growth [159].

3.4.3.4   Creating biotopes

In recent years, wetland biotopes have been created as compensation measures on surrounding land by taking advantage of the topography at each site. At the Kanna Dam (PG, 45.0 m, #136), low weirs are constructed on tributaries feeding the reservoir to form a wetland zone where aquatic plants can grow. Downstream from the dam,

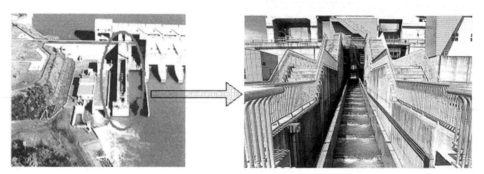

Presented by MLIT

Figure 3.4.3 Fishway of Nibutani Dam. (See colour plate section)

Presented by MLIT

Figure 3.4.4 Mangrove forests downstream from the Kanna Dam. (See colour plate section)

advantage was taken of the river's tidal characteristics to create a mangrove forest, which is a distinctive feature of river mouths in Okinawa (Figure 3.4.4).

At the Miyagase Dam, a pond and mound were formed by guiding surface water on a dumping yard (an area of 11 ha) located in a mountain torrent, thus creating a biotope. Monitoring has obtained very interesting data, confirming that over time, a variety of forms of life came to inhabit the biotope and that it was used as a watering site by deer and wild boar.

3.4.3.5 Flexible reservoir management [160]

It has recently been pointed out that dams, by creating extremely smooth flow regimes downstream, have an impact on the natural environments of downstream rivers. Therefore, a Test for Flexible Reservoir Management to ensure better environments by carrying out operations in a flexible manner, is performed at dams under the direct management of the MLIT.

Flexible management means adjusting the water level for flexible use at an elevation that is higher than the stipulated water level, such as the normal maximum water level when there is no danger of flood, and storing water in this newly created capacity to increase the quantity of the river maintenance flow rate discharged or to execute flushing discharge[10] etc. Tests for Flexible Reservoir Management were planned for 20 dams in 2001.

Improvements in river environments downstream from these dams were verified as described below.

3.4.3.5.1 The effectiveness of cleaning riverbed deposits with flushing discharge

At the Sagae Dam, flushing discharge to remove silt deposited on the riverbed was done 16 times at 10, 20, and 30 m^3/s, while the usual river maintenance flow rate is 1.2 m^3/s at a downstream control point. This flushing removed the silt that had been deposited on the riverbed.

3.4.3.5.2 The effectiveness of flushing discharge as a stagnant water countermeasure

At the Miharu Dam, flushing discharge was done 11 times at 20 m^3/s, while the usual river maintenance flow rate is 1.2 m^3/s at a downstream control point, thereby washing out stagnant water that causes bad odors.

3.4.3.5.3 The effectiveness of an increased river maintenance flow rate as a means of improving river scenery

At the Izarigawa Dam (ER, 45.5 m #4), 0.3 m^3/s of water was discharged into a dried-up lower reach of the river for 39 days, effectively improving the river scenery.

3.4.3.5.4 The effectiveness of increasing the river maintenance flow rate on fish habitats

At the Managawa Dam (VA, 127.5 m, #61), discharge was increased by 1.0 m^3/s for a total of 16 days, while the river maintenance flow rate is 0.28 m^3/s, resulting in an increase in the percentage of rapids as shown in Figure 3.4.5. Although no change of

---

10 Flushing discharge: when the flow rate remains constantly low for a long period of time, algae, dirt, etc. adhering to the riverbed remain intact, creating stagnant water, so the periodic discharge of stored water from the dam alters the flow regime, thus suppressing stagnation and refreshing the river.

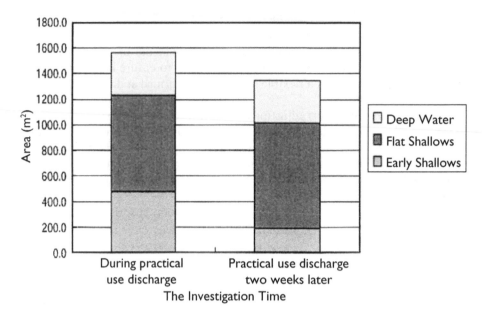

Figure 3.4.5 The changing area of rapids and pools during and after flexible use discharge at the Managawa Dam.

Source: T. Osugi, M. Urakami, Report of Water Resources Environment Research Institute, 2002 [160].

the number of sweetfish was observed directly below the dam, traces of feeding were confirmed over wider areas during the discharge period.

The reservoir capacity for flexible reservoir management was limited because of restrictions on flood control, so it is necessary to carry out a thorough preliminary study to verify the effectiveness thereof, based on objective evaluation indices, before actual flushing discharge commences.

3.5   A REFLECTION OF THE VIEWS OF CITIZENS CONCERNING DAM PROJECTS AND AN INTRODUCTION TO THE PROJECT EVALUATION SYSTEM

In recent years in Japan, there has been frequent suspicion that large-scale public works projects such as dam construction projects are wasteful, due to the increasing length of time they require, their soaring costs and their environmental impacts. Such suspicion is often caused by a lack of objective project evaluations and insufficient transparency of the project decision-making and revision processes.

This section describes the history and background of efforts made in response to this problem to collect and reflect the views of local residents and experts.

3.5.1 The escalating citizens' movements against dam projects [161–163]

3.5.1.1 Dealing with the movement against the construction of the Nagaragawa Estuary Barrage

The Nagaragawa Estuary Barrage (Weir, #86, Figure 3.5.1) was constructed to prevent saltwater intrusion. This was accompanied by dredging of the river course to expand the flood discharge capacity of the Nagara River. The plan also included providing new supplies of municipal and industrial water. This project stimulated people's concern regarding the advisability of its construction. Accordingly, the river manager (the MLIT) and the implementing body (the JWA), published accurate information about the project and actively participated in discussions with local residents.

Specifically, from 1992 until 1994 before the barrage started operating, the MLIT and the JWA discussed the issue with citizens groups. Later total of eight round-table discussions in public forums between supporters and opponents of the estuary barrage were held from March to April 1995. Participants included academics, municipal

Presented by JWA

Figure 3.5.1 Nagaragawa Estuary Barrage. (See colour plate section)

officials, residents holding pro- and anti-project views, the MLIT and the JWA. The topics comprised four themes: disaster prevention, salt damage, the environment and water demand and supply. Following this process, full-scale operation of the estuary barrage began in July 1995.

After operation began, "New Dialogues", which are discussions between citizens and the MLIT and the JWA, were conducted in 1995 and 1996. The Nagaragawa Estuary Barrage Monitoring Committee of academic experts was formed in 1995, and with its guidance and advice, monitoring concerning disaster prevention, water quality, bottom sediment and ecosystems commenced. The monitoring was announced publicly, extending from its implementation to its results, and the committee started holding public discussions in 1997. The committee presented its monitoring plan for the Nagaragawa Estuary Barrage to the MLIT and the JWA in March 2003, then disbanded, and monitoring has continued based on the original plan.

3.5.1.2 Establishing the evaluation council for dam projects

The experience gained from the Nagaragawa Estuary Barrage conflict has significantly influenced the execution of subsequent large-scale public works projects in Japan.

Assuming that it is necessary to provide procedures for hearing the views of local residents in order to ensure the transparency and objectivity of dam projects, in July 1995 when the estuary barrage began full operation, the Evaluation Council for Dam Projects was established to discuss the purpose and contents of every project to construct dams, weirs, etc. by the MLIT and JWA. The members of the Evaluation Council are selected on recommendation by the concerned prefectural governor, who is the representative of the region, in order to hear the genuine views of the region. The Evaluation Council heard a wide range of views from local residents and nature conservation groups by holding public hearings as necessary, and established a specialized survey committee of experts. After deliberations, the Council summarized the views of project policies (continuation, revision, abandonment, etc.) and submitted these to the body implementing the project. Under this system, Evaluation Councils have been established to deal with 14 projects and have submitted their views on each project.

3.5.1.3 Introduction of the monitoring and evaluation
system for dam management

The Evaluation Council for Dam Projects has taken the initiative with projects during construction, but in order to achieve even more appropriate management by accurately clarifying the effectiveness of completed dams and those already in operation and to assess their environmental impacts, in 1996 it began a trial of the Monitoring and Evaluation System for Dam Management for dams and weirs operated by the MLIT and the JWA. This system began full operation in 2002. Based on this system, committees of academic experts are formed for each dam, etc. to carry out periodic analysis and evaluation of the results of surveys of flood control by the dam, and its impact on the environment, etc. These committees release reports of their results to the public.

3.5.2 The reflection of citizens' views of dam projects

3.5.2.1 Revision of the River Law to reflect views of residents

The River Law is the basic law regarding flood control projects and it formerly stipulated that when a river manager enacted a Basic Plan for the Implementation of Construction Works for a river system, the river manager must hear the views of the River Council, but it did not provide for procedures to reflect the views of residents.

On the other hand, the needs of citizens have diversified to include the improvement and conservation of river environments, so it is clear that links with local residents are indispensable to improving rivers according to the characteristics of the river, its natural features and culture of the region. It was, therefore, necessary to provide procedures to reflect the views of local governments and regional residents in the enactment of plans to improve rivers.

Under these circumstances, a significant revision of the River Law was carried out in 1997. The major changes are as follows:

a "Improvement and conservation of river environments" was added to the purposes of the River Law.
b It prescribes that in place of the former Basic Plan for the Implementation of Works, a Fundamental River Management Policy, which is a long-term plan and a River Improvement Plan[11], which is a plan covering approximately 30 years, must be enacted.
c It also prescribes that when a River Improvement Plan is enacted, as necessary, measures needed to hear the views of academic experts and to reflect the views of concerned residents must be taken.

After the revision of the River Law, procedures to reflect the views of residents in a dam project were stipulated by law.

3.5.2.2 Examples of the application of the new River Law to dam projects

Under the revised River Law of 1997, a river basin committee is formed for each river to hear the views of local governments, academic experts and river basin residents in preparation for drawing up a River Improvement Plan. It deliberates on specific

---

11 River Improvement Plan: Article 16 of the River Law stipulates items related to specific river works, such as dams, levees, etc. and the contents of river maintenance plans in districts where river managers must carry out river management systematically, in line with the Fundamental River Management Policy.

improvements to the river from a broad perspective, including flood control, water usage and the environment. By the end of the fiscal 2004, River Improvement Plans had been prepared for 15 Class A River Systems. Of these, dam projects that had become the center of major disputes were the Toyo River System Improvement Project (Shitara Dam) and the Hiji River System Improvement Project (Yamatosaka Dam (PG, 103.0 m, #115)).

On the Toyo River System, a river basin committee was formed in December 1998 and they met a total of 23 times until the River Improvement Plan was completed in October 2001. It positioned the Shitara Dam (PG, 129.0 m, #85) as a facility required for flood control, water supply and environmental improvement in the river basin.

On the Hiji River System, a river basin committee was formed on October 2003 and had met four times by March 2004. During this period, the committee collected opinions using postcards and the Internet, held meetings to exchange views with residents, and publicly disclosed the contents of the improvement plans on cable television to gather the opinions of river basin residents. Then, as a result of deliberations based on these activities, it announced the Hiji River Improvement Plan in May 2004.

To study this River Improvement Plan, it decided what form the river basin and river should have, selected and set numerous proposals for ways to achieve this ideal form, and tested methods of comparing these multiple proposals, including the results of EIA [164].

3.5.3   Introduction of the project evaluation system [165]

3.5.3.1   Improvement of the public works project evaluation system and the introduction of the policy evaluation system

As stated above, project evaluation of dam projects by the Evaluation Council for Dam Projects had been introduced in advance, then the evaluation of road projects, wastewater system projects and other public works projects under the jurisdiction of the MLIT started in 1997. As shown by the example of the trial of cost-benefit analysis at the project budget adoption stage for large-scale public works projects introduced in 1997 by the then Ministry of Transport, which is in charge of port and harbor projects, efforts to objectively evaluate the need for, and effectiveness of, the implementation of projects have been spreading gradually.

In December 1997, the Prime Minister instructed six ministries and agencies that execute public work projects to introduce a public works re-evaluation system and to perform cost-benefit analyses at the budget adoption stage. Each ministry and agency decided to begin implementing these tasks in 1998.

The MLIT, which was in charge of dam projects, began performing re-evaluations of all public works projects under its jurisdiction in fiscal 1998, with the Committee for Project Evaluation and Monitoring deliberating on each project.

The MAFF also introduced a re-evaluation system for agricultural dam construction projects.

However, this did not stop public criticism of public works projects, so in August 2000, the ruling parties at that time requested a radical review of public works projects that satisfied the following four standards:

a   Projects where work has not begun within five years of adoption.
b   Projects that have not been completed within 20 years of adoption.
c   Projects that are presently suspended (frozen).
d   Projects that have not been adopted within 10 years of starting their planning survey for execution.

The MLIT performed an independent review to add additional standards to these four standards, and in response to deliberations by the Committee for Project Evaluation and Monitoring on 48 dam projects that satisfied both standards (out of a total of 136 projects under the jurisdiction of the MLIT), they abandoned work on 46 dam projects.

Later in July 2001, the MLIT enacted and began enforcing unified rules for evaluations when adopting a new project, and for re-evaluating public works projects under its jurisdiction, including dam projects.

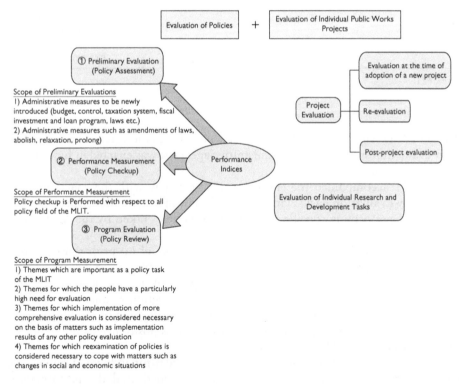

Figure 3.5.2 Evaluation system based on the MLIT Policy Evaluation Basic Plan.

Source: MLIT.

Table 3.5.1 The present project evaluation system of the MLIT (Outline).

	Year of introduction (fiscal year)	Time when the evaluation is executed	Outline
Evaluation at the time of adoption of a new project	1998	When the budget for a new project is adopted.	The evaluation including the cost benefit analysis is executed.
Re-evaluation	1998	Five years have passed since the project was adopted, and it does not start con-struction. Ten years pass and the ongoing project. etc.	The project is reviewed if necessary, or the project is stopped when not admitted that its continuance is suitable.
Post-project evaluation	2003	Within five years after completing the project.	The effect and the envi-ronmental impact of the project are confirmed. If it is necessary, an appropriate measure for improvement or the ideal way of the plan and examination of the similar project are examined.

Source: MLIT.

At about the same time, the Government Policy Evaluation Act, which stipulates a policy evaluation system for all administrative agencies, was enacted in June 2001 and came into force in April 2002.

The MLIT accompanied the enforcement of this law with the enactment of the MLIT Policy Evaluation Basic Plan based on the law in March 2002, thus establishing a method of performing specific evaluations of policies taken by the MLIT (Figure 3.5.2).

Under this Basic Plan, a policy evaluation consisting of (a) prior evaluations that assess the need for, effectiveness of, and efficiency of new policies (policy assessments), (b) performance measurements to assess the degree of achievement of previously-stipulated policy goals (policy check-ups), and (c) program evaluations to discover challenges and improvement measures based on the verification of the effectiveness of implementing each type of policy (policy reviews), was positioned as the basic policy evaluation to be applied by the MLIT.

Under this Basic Plan, the project evaluation system for individual public works projects that had been used in the past was expanded, and the performance of three evaluations – (a) evaluation at the time of adoption of a new project, (b) re-evaluations, and (c) evaluations performed after a project has been completed (post-project evaluations) – was clearly stipulated. Evaluations are now done according to implementation rules that have established specific methods for performing each type of evaluation (Table 3.5.1).

Post-project evaluations of individual dam projects are, having already been systematized by the Monitoring and Evaluation System for Dam Management (see 3.5.1.3), evaluations based on this system.

The MAFF has also enacted the Policy Evaluation Basic Plan and the Policy Evaluation Implementation Plan, and carries out re-evaluations of past projects, prior evaluations performed when starting new construction, and post-project evaluations performed five years after completion of a project.

3.5.3.2   Policy reviews of dam projects, etc. (program evaluations)

3.5.3.2.1   Dam projects – verification of effects and impacts on a region – [166]

Under the MLIT Policy Evaluation Basic Plan, policy reviews (program evaluations) of policies that are extremely important as policy challenges, policies whose evaluations are demanded by people, and policies that require revision in response to changed social and economic conditions have been performed. So the dam project policy was selected as one of the first policies to be evaluated after enactment of the planning system, and a policy review was carried out from 2001 to 2002. As a result, in March 2003, it was prepared as the Dam Projects – Verification of Effects and Impacts on a Region – and released to the public, so it is introduced here.

This policy review clarified the purposes of dam projects as, "Achieving the purposes stipulated in Article 1 of the River Law, which are (a) preventing the occurrence of disasters caused by floods, etc. (b) using rivers appropriately, (c) maintaining the normal function of river water, and (d) improving and conserving river environments," and evaluations are done from the following perspectives:

1   Was the dam project able to control floods, supply water and achieve its other anticipated goals?
2   Including the manner in which a project was conducted, what measures have been taken to deal with challenges, such as its impact on regional societies, the natural environment, the water cycle, etc.?
3   Directions of efficient and effective dam project improvements, based on changes, etc. in social and economic conditions in recent years.

The results have clarified that flood damage has been reduced and water supplies have been stabilized by dams operated by the central government and by the JWA.

On the other hand, the disappearance or contraction of hamlets, the division of regions and other impacts on regional society caused by constructing dams, plus the impact on natural environments by dam projects that form reservoirs, build diversion roads, etc. and other large-scale changes to the topography were categorized. And basic directions concerning measures such as supporting the re-establishment of the livelihoods of relocated people and revitalizing the reservoir region by a water resource region improvement project, evaluating environmental impacts and conserving environments, are also clearly described, together with past countermeasures taken to deal with these impacts. In addition, items such as the following that should be introduced in the future are presented.

1   Setting efficient operating rules to achieve staged flood control targets with maximum effectiveness.
2   Thorough utilization of existing facilities by, for example, reorganizing capacity to optimize the functionality of a group of existing dams.
3   The study of securing the stability of water supply under the trend of declining precipitation and the best way to share burdens.

3.5.3.2.2   Water usage adjustment to improve river environments – improvement of dried-up rivers caused by water intake [167]

As a dam project related policy review, "Water usage adjustment to improve a river environment – improvement of dried-up rivers caused by water intake", conducted from 2001 to 2002, was evaluated.

In Japan, in order to restore abundant flowing water in rivers and resolve the social problems caused by the drying up of rivers and the decline of the water environment in various regions as a result of using water to produce hydropower. In 1988, electric power companies and river managers reached an agreement to maintain a constant river flow when water rights for hydropower were periodically renewed. Then, after more than ten years had passed, the river maintenance flow had been improved at many places, so this policy was selected as the object of evaluation in order to verify how effectively it restored clear flow and to confirm the evaluation of local people.

This policy review was a questionnaire survey directed at river managers and local governments, and a field survey at model dams, which clarified and analyzed the river environment improvement achieved by discharge of the river maintenance flow, in order to answer two questions about the discharge of the maintenance flow that was carried out when water rights were renewed: (a) how is the discharge of the river maintenance flow evaluated in the region where it has been done, and (b) what impact has the discharge of the river maintenance flow had on the river environment?

Because positive evaluations have been obtained concerning the discharge of a river maintenance flow, the MLIT continues to establish appropriate river maintenance flow and also, if possible, studies effective discharge methods such as varying the discharge rate by season or performing flush discharge, etc. and strives to improve the flow regime.

3.5.3.3   The importance of project evaluations at dam projects and future challenges (project cost management and technological development)

Those implementing past dam projects strove to ensure the benefits of investment in the projects by promoting effective and efficient projects and early manifestation of their effects. However, under recent harsh financial circumstances, the central and local governments have had to restrict expenditure on public works projects. During a dam project, the effectiveness of the investment is strictly evaluated at each stage [168] by a prior evaluation, re-evaluation, and post-project evaluation, using the project evaluation system described above.

Therefore, efforts have been made to develop new technologies in the areas of materials, design and execution, typified by the RCD (Roller Compacted Dam Concrete) method and trapezoidal CSG (Cemented Sand and Gravel) dams, and to cut costs by using these new technologies. In recent years, new contracting systems with the priority on technology and price were also introduced and these are gradually taking effect. These efforts must be strengthened in the future.

Furthermore, it must not be forgotten that ways of cutting costs include the concepts of the lifecycle cost that comprises measures to improve the durability of facilities and extend their lifespan.

Chapter 4

Roles of dams: The future

Chapter 2 describes the roles that dams have played in the past and Chapter 3 examines the negative impacts of dams that are now in operation and measures taken in response to these impacts. This chapter looks to the future to describe the roles that dams should and can play, and the tasks that dam engineers should perform from an international perspective when looking into the future.

Dams have a past, a present, and also a future. Needless to say, in the future it will still be important for dams to play their existing roles reliably and for great efforts to be made to reduce their negative impacts. However, future changes in social conditions will probably transform the roles of dams.

Existing dams must be well maintained and their effective use must be promoted in order that they continue to play their roles reliably in the future, and engineers must support the achievement of these goals through study and hard work. This technology will be of great use in various parts of the world.

4.1 CHANGES IN THE SOCIAL CONDITIONS SURROUNDING DAMS

4.1.1 World trends

4.1.1.1 The growing world population

The United Nations estimates that the world's population will reach 7.9 billion by 2025, an increase of 1.9 billion people from 2000. By 2050, the world population is predicted to reach 9.3 billion, a further increase of 1.4 billion people (Figure 4.1.1) [169]. This total increase of 3.3 billion people will include an increase in developing countries, excluding China. There is concern that this population increase will further widen gaps between supply and demand for water and food, which are essential for life, thereby causing serious international problems. It is also predicted that the world's population will approach 10 billion in the latter half of the 21st century.

4.1.1.2 Intensification of food shortages

International supply and demand for agricultural products is already unstable: if, during steadily rising consumption, crop failures caused by unseasonable weather

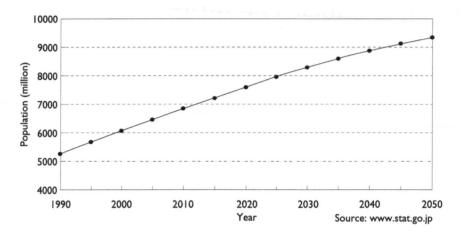

Figure 4.1.1 Predictions of world population.

Source: Ministry of Internal Affairs and Communications [169].

occur in major producing countries such as the U.S.A., production will fall far short of consumption, resulting in soaring prices. Recent abnormal weather patterns have increased the potential for fluctuations in food yields and quality.

It is predicted that in the medium to long term, rising income in developing countries in conjunction with the rise in the world's population will expand the consumption of livestock in Asia, thereby causing a sharp rise in demand for feed grains. On the supply side, increased productivity has been supported by greater yields per unit area by improving breeds, introducing chemical fertilizers and constructing irrigation systems. However, there is clearly a variety of factors that restrict productivity, such as restrictions on expansion in the area of cultivated land and environmental problems [170].

According to a prediction by the UN Food and Agricultural Organization (FAO), in order to respond to growing grain demand, world grain production will rise by about 500 million tons between 2000 and 2025 (Figure 4.1.2) [171]. According to the FAO, by 2030, it will be necessary to develop 120 million hectares of new agricultural land [170].

Future expansion of cultivated land must be done by developing land with low productivity that is less fertile and difficult to irrigate. On the other hand, the overgrazing of grassland, salt accumulation caused by excessive pumping of groundwater, and the failure of irrigation etc. have caused desertification at a rate of 6 million hectares per year [172].

The international interaction of the various factors described above has increased the need for the development of new water resources to supply irrigation water.

4.1.1.3 A shortage of water supply and access to sanitary water

It is reported that about 8% of the world's population suffer from a shortage of water, despite living in regions where large quantities of water are already being used. It is also reported that in 2000, 1.1 billon people, or about 20% of the world's

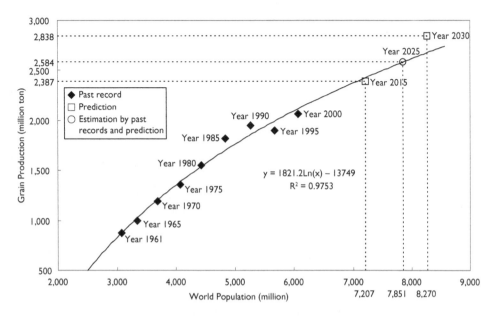

Figure 4.1.2 Predictions of the future world population and grain production [172].

Source: Water Resources Dept., MLIT.

population, did not have access to a supply of safe water [173]. It is predicted that in 49 developing countries, including Bangladesh and Nigeria where the population growth rate is as high as 2.4%/year [174], access to sanitary water will become even more difficult.

In a number of regions, the decline in river flow rates, falling water levels in lakes and marshes, ground subsidence and the influence of saline water caused by excessive pumping of groundwater have already caused harm. In the future, these harmful trends will worsen.

As urbanization advances worldwide, it is becoming vital to ensure supplies of drinking water in response to rising regional demand.

4.1.1.4 Energy problems

The quantity of energy consumed worldwide continues to grow steadily, with more than half of this being consumed in North America, Europe and other industrialized countries. Nevertheless, the rate of increase is growing relatively slowly. While the quantity of energy consumed in developing countries is currently low, the rate of increase will continue to rise sharply as a result of population growth and industrialization in China and other countries in Asia [175].

The sources of energy being consumed are mainly oil, coal, natural gas and other fossil fuels. It is claimed that while there are approximately 200 years of coal reserves, there are approximately 50 years of reserves of other energy sources like oil, natural gas and uranium. Therefore, in the future it will be important to develop substitute sources of energy [176].

The example of the Three Gorges Dam in China, along with global warming caused by using fossil energy, shows that hydropower production will probably be reconsidered as a valuable source of energy in countries where there are still good development locations.

4.1.1.5   Global warming

CO_2 and other greenhouse gases are the major cause of global warming, resulting in climatic fluctuations that raise fears of rising sea levels, more frequent floods and droughts, a decline in agricultural yields, impacts on ecosystems, etc. It is predicted that if regulations are not enforced, by the end of the 21st century the average temperature of the earth will have risen by 1°C to 3.5°C, thereby raising the sea level by several dozen centimeters. It is forecast that the temperature will increase most in the higher latitudes of the northern hemisphere and least in Antarctica and the northern regions of the northern Pacific Ocean. Also, because of the intensification of evaporation and other phases in the water cycle between the atmosphere and the ocean, rainfall and the quantity of moisture in soil will rise in the northern latitudes in winter [177].

4.1.2   The future of Japan related to dams

4.1.2.1   Population decline and the urban–rural gap

The total population of Japan in 2005 was 127.77 million people. According to the results of one estimate in 2006 [178], the population peaked in 2006, and is predicted to be 115 million by 2030, which is not very different from the present population. However, it is predicted to decrease to 99 million in 2046 and 90 million by 2055, representing a fall of 70% from present population levels.

By 2030, the population will be unchanged from the 2000 level in Saitama, Chiba, and Kanagawa Prefectures that border on Tokyo, and in Miyagi, Aichi, and Fukuoka Prefectures that include large regional cities. However, it is estimated that in Akita, Yamaguchi, Nagasaki Prefectures, etc. the population will plunge and that the percentage of elderly people in the total population will rise [179]. Accompanied by an overall declining population, this will result in a relative concentration in cities and the depopulation of rural regions. Rapid decrease in population of the rural regions is predicted to harm the environment by lowering the levels of forest and farmland management. It will also hinder the passing down of social, cultural and natural values to coming generations.

4.1.2.2   Agriculture and agricultural water

It is predicted that the population of Japan will remain at its present level until about 2025 then it will decline between 2050 and 2100. Within Japan, the demand for irrigation water, including that provided by irrigation dams, is unlikely to increase.

Nevertheless, agricultural products are initially produced for consumption within the country of production, with only a surplus being made available for export. Agricultural products cannot be stored for as long as industrial products without deteriorating. For these reasons, agricultural products account for a small percentage of those produced for, and handled by, international trade.

Japan's self-sufficiency in food is about 40% on a calorie-intake basis, which is extremely low for an industrialized country. This means that Japan will still be short of food, even after its population has halved in size. A more detailed examination of this self-sufficiency rate shows that it is 95% for rice, 83% for potatoes and 82% for vegetables, which means that its self-sufficiency rate for foods needed to maintain the traditional Japanese diet is relatively high, but its self-sufficiency rate for meat is 54%, so it imports product that are indispensable in the Western diet. However, as Japan's self-sufficiency rate for feed to raise livestock is low at 24%, its self-sufficiency rate for meat is actually even lower: closer to 40% on a calorie-intake basis [180]. Therefore, if the world's population continues to rise, causing even worse food shortages and global warming problems, the pressure on Japan to be self-sufficient in food production will increase.

On the other hand, water problems caused by imported foods have been identified. For example, it is reported that it takes 2,000 liters of water to produce either 1 kilogram of wheat or 100 grams of beef. The water necessary to produce imported grain, meat, etc. in this way is called virtual water. It is estimated that the total water resources, or in other words the virtual water, required to produce grain or meat, etc. imported by Japan in 2000, was 64 billion m^3/year [181]. This quantity is quite a large amount of water, considering the fact that the total quantity of water used in Japan (on an intake basis) is 85.2 billion m^3/year. Agricultural production cannot respond quickly to changes in supply and demand, because production requires a fixed period. It will therefore be necessary to account for the need to develop new water resources, considering the surplus water created by the inability to use rice production facilities because of a falling population and change in consumption trends, along with rising demand for other types of agricultural production.

4.1.2.3 Energy and hydropower

Just before the first oil shock, oil provided the highest percentage of Japan's primary energy, accounting for 77% of all energy. Later, the oil shocks led to the introduction of nuclear power, liquefied natural gas (LNG), and coal, etc., so that by 1998, Japan's dependency on oil was down to about 52% [182].

The percentage of primary energy provided as electric power was 41% in 2000, while other forms were sent directly to consumers as fuel [183]. With the exception of the small supplies provided by new energy sources (about 1%), energy, other than electricity, is produced almost entirely from fossil fuels. To prepare for the depletion of fossil fuels and to stop global warming, hydrogen and other secondary energy media should be developed. If hydrogen energy becomes a replacement fuel for oil energy in the near future, electric energy will be needed to produce hydrogen, and in order for this to be as independent of fossil fuels as possible, nuclear power, new energies and hydropower must be developed.

Next, an examination of the breakdown of electric power sources shows that it used to be mainly hydropower, but from about 1962, hydropower was surpassed by thermal power. Fuels used to produce thermal power are oil, coal, LNG, etc. At peak production

Table 4.1.1 Comparison of electric power sources.

Electric power sources	Ratio of generated energy (as of end of FY 2002)	Generation cost (Yen/kWh)	Unit CO_2 discharge (g-CO_2/kWh)
Hydro	9	13.6	11
Nuclear	31	5.9	22
Coal-fired thermal	22	6.5	975
LNG-fired thermal	27	6.4	500 to 600
Oil-fired thermal	9	10.2	742
New energy			
Wind	<1 (Total of new energy)	10 to 24	29
Waste material		9 to 12	

Source: METI.

times, more than 60% of all electric power was produced from oil. The oil shocks were followed by the development of electric power sources such as nuclear power, coal, LNG, etc. as substitutes for the oil that is high-priced and its supply is unstable [184].

As shown in Table 4.1.1, nuclear power now accounts for 31% of annual electric power production. Incidentally, hydropower accounts for 9% of annual electric power production.

The share of oil used for electric power production is low at 9%. Its electric power production cost is high and it generates a lot of CO_2. Therefore, the use of oil will continue to decline in the future.

Nuclear power, which is the largest source, generates almost no CO_2 and its production cost is the lowest, but recent accidents at nuclear power plants have made it difficult to boost nuclear power, and ensuring safety and back-end measures[1] are other challenges.

The second-largest source, LNG, provides relatively low-cost power and is considered the cleanest among fossil fuel sources, but it produces a lot of CO_2: between 500 and 600 g/kWh.

Coal, which is the third largest source of electric power, ensures superior fuel supply stability and cost, but it produces 975 g/kWh of CO_2, the largest of any energy source.

The fourth source is hydropower, which produces extremely small quantities of CO_2, even less than new energy sources, is clean, and as energy produced entirely in Japan, offers a very stable supply. Although hydropower's initial investment costs are relatively high, its long-term cost is low. To develop future hydropower plants, it is necessary to reduce costs and protect the environment. In addition, it is important that efficient maintenance be performed continuously to extend the equipment's lifetime in existing hydropower systems, which offer long-term cost superiority.

1 Back-end measures: to promote the development and use of nuclear power, it is important to establish measures to treat and dispose of the radioactive waste that nuclear power produces and to close down and dispose of nuclear power facilities. These are called nuclear power back-end measures.

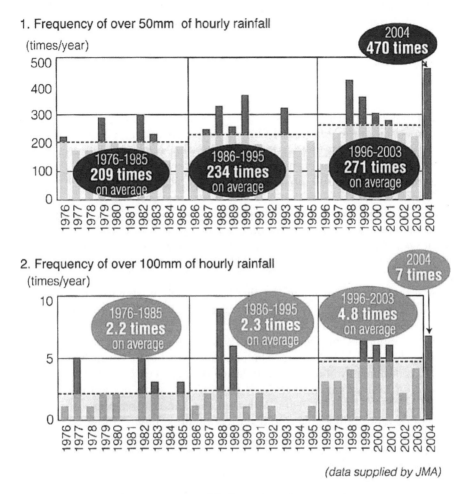

1. Frequency of over 50mm of hourly rainfall

2. Frequency of over 100mm of hourly rainfall

(data supplied by JMA)

Figure 4.1.3 Frequency of concentrated rainfall. (See colour plate section)
Source: MLIT [186].

New energies still provide less than 1% of electric power generation, but under the RPS Act[2], electric power companies are now legally required to provide a certain percentage of their power by wind power, waste material power, or other reusable energies, and this is expected to promote the development and spread of these

2 RPS Act: Act on Special Measures concerning New Energy Use by Electric Utilities (Renewables Portfolio Standard) (2002). This law obligates electric power companies to use new energy to supply a stipulated percentage, or more, of their electricity according to the quantity of electricity it sells in order to promote the use of new energy sources. The types of energy covered are photovoltaic power, wind power, biomass, geothermal, small hydropower, etc. Small hydropower refers to conduit type hydropower plants with an intake weir with height shorter than 15 m and to plants producing an output of 1,000 kW or less, using the maintenance flow from dams.

technologies. Nevertheless, to develop these in the future, their costs must be reduced and they must supply electricity more stably.

4.1.2.4 Predicted abnormal weather

According to the latest understanding of global warming, abnormally high air temperatures will occur more frequently, abnormally low temperatures more infrequently, and rainfall will be either extremely high or extremely low [185].

A comparison of the frequency of hourly rainfall of 50 mm or more and 100 mm or more in Japan during the last 30 years, in 10-year blocks, shows a staged increase in the frequency of both phenomena, as shown in Figure 4.1.3. The rise in the frequency of hourly 100 mm or higher rainfall seen in recent years is particularly conspicuous.

While an annual average of 2.6 typhoons hit Japan, in 2004, 10 typhoons, or about four times as many as usual, did so, causing severe flood damage. As shown in Figure 4.1.4 all dams in Japan regulate flooding 396 times/year on average, but in 2004, they performed flood control 933 times, which is about 2.4 times the usual frequency.

Annual precipitation in Japan has fluctuated widely in recent years, as shown in Figure 1.4.2 in Chapter 1, setting a new record low for annual precipitation. Examining trends shows that during the past 100 years, the average annual rainfall has declined by about 100 mm from that of 1900 [188]. There are reports of other data indicating that in Tohoku, Kanto, Shikoku, and Northern-Kyushu Regions, annual rainfall has declined by about 200 mm during the same century [189].

According to a recent prediction of how global warming will change the climate during the next 100 years, although the annual precipitation will go up, there will be more rainless days and snowfall will go down. Snowfall will decline particularly sharply in the area from Hokkaido to the Chugoku Region along the Japan Sea.

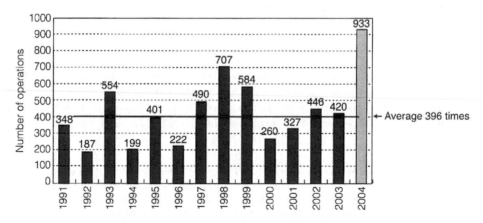

Figure 4.1.4 Frequency of flood control operations (1991/2004).

Source: MLIT [187].

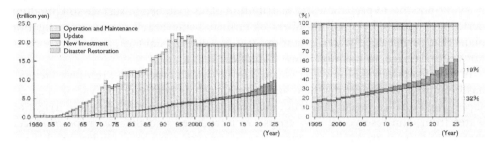

Figure 4.1.5 Estimation of investment demand for maintenance and renewal of public infrastructure. (See colour plate section)

Source: MLIT [190].

4.1.2.5 Increase in maintenance costs of public infrastructure

Because many infrastructures constructed during the rapid economic growth period around the 1960s are deteriorating as they age, the cost of maintaining existing stock is likely to rise. It is also foreseen that even if the growth of public works investment is assumed to be zero, by about 2025 maintenance and renewal investment costs will exceed new investment and account for more than 50% of public works investments. As the falling birth rate and aging society are pushing up the cost of social welfare and suppressing public works investments, the percentage of public works investments devoted to maintenance and renewal will rise (Figure 4.1.5) [190].

It is therefore necessary to maintain existing dams in good condition to prolong their lives and to effectively use them by redeveloping them. It is also necessary to reduce the costs of dam construction, considering their life cycle costs.

4.2 ROLES OF FUTURE DAMS

4.2.1 Agriculture and dams

Ensuring stable supplies of safe food will continue to be indispensable to enable future generations to live their lives without concern. It is therefore necessary to maintain agricultural production in Japan. However, agricultural production is declining, and although it is governed by peoples' diets, self-sufficiency in food is the lowest among the major industrialized countries.

To increase domestic agricultural production and ensure stable supplies of food in the future, efficient and stable farm operations must support a substantial portion of domestic production.

A stable supply of irrigation water must be ensured, not only to provide a supplementary supply during times of drought, but also to perform high-productivity paddy field agriculture during normal times. Irrigation ponds and dams that control discharge in annual units by storing unstably flowing water in order to distribute the necessary quantities at necessary times and at low cost to paddy fields must be effective, as a prerequisite for agricultural management. The roles and functions of existing dams and irrigation ponds must be strengthened and maintained so that in the future, it will be possible to prepare water distribution plans matched to surface soil puddling performed quickly by small numbers of farmers using heavy machinery, irrigation plans that permit deep water irrigation[3] to prevent frost damage, and to respond to hot water damage, to mention a few examples.

In recent years, irrigation water ensured by dams and other water resource facilities has been used to support dry-field agriculture, such as hothouse cultivation that can supply high-quality vegetables, fruit, etc. throughout the year. With stable irrigation water from dam reservoirs, new crops are introduced and diverse crops produced throughout the year. More advanced use of irrigation water provided by dams is anticipated as a way to achieve the improvement of environments that nurture crops and farm workers' working environments. Moreover, in the future when the pressure to improve Japan's self-sufficiency in food will probably rise, it will continue to be important to effectively use existing dams in order to ensure irrigation water, thus contributing to the stable supply of food during serious droughts. This will require that dams be operated and managed to supply water and that dam and reservoir safety management be easily understood, so it can be carried out properly.

4.2.2 Energy and dams

The following will be important as policies to restrict emissions of CO_2, thereby lowering dependency on fossil fuels while responding to the anticipated increase in worldwide energy consumption:

a   Restricting energy consumption and saving energy in industrialized countries.
b   Developing and adopting recyclable energy (hydropower, wind power, geothermal, photovoltaic, hydrogen, wave power, seawater temperature-difference power, bio-fuel, etc.).

The benefits of hydropower production, which is one type of recyclable energy, are that it is a purely domestic recyclable resource that can be recovered, produces low emissions of CO_2, has a long service life, contributes to regional development, and its technologies are established.

The CO_2 reduction effects of a hydropower plant can be assessed according to the area of forest required to absorb the CO_2 produced, assuming that the same power was produced by oil-powered thermal power.

Its negative impacts are that it is expensive and that it may affect the natural environment.

Japan produces $1,076 \times 10^6$ MWh of electric power, of which 94×10^6 MWh is hydropower [191]. Including existing hydropower, Japan's potential hydroelectricity equals 135×10^6 MWh [192].

It is presumed that hydropower will be implemented as described below.

---

3 Deep water irrigation: This is a method of irrigating at increased depth in order to protect young panicles from the cold air during the frost season when paddy rice is susceptible to damage from low temperatures.

At this time, construction at almost all locations suitable for large-scale development has been completed, and in this century, it will be necessary to maintain, manage, and prolong the life of dams and hydropower plants that have already been constructed. Future development will presumably be done by introducing power production technologies that are kind to the environment and suited to locations with short falls and low flow rates, thereby developing hydropower plants that take advantage of unused small falls, including that at dams other than hydropower dams, while reducing production costs. Hydropower is a clean 100% domestically produced recyclable energy, and will naturally be passed on to future generations as a valuable asset.

Medium and small hydropower development will make a great contribution to ensuring valuable domestically produced energy, and will go beyond merely producing power to create regional industries centered on hydropower generation, etc., by providing other functions that contribute to autonomous development of the region. A trend towards using small falls effectively, mainly on agricultural channels and rivers at a scale ranging from a few tens of kilowatts to a few hundred kilowatts, has emerged.

Hydropower generation during this century will probably meet the demands of the times from various perspectives, including resolving global environmental problems.

4.2.3 Abnormal weather and dams

Even in Japan, abnormally heavy rain, abnormal drought, unusually high and low temperatures, and similar phenomena believed to be caused by global warming have been observed. Both record-setting heavy rain and localized concentrated torrential rain have occurred frequently throughout Japan. In the future, such changes in the weather are likely to increase in intensity, thus having a greater impact on people's lives along with a great environmental impact on fauna and flora dependent on rivers.

The present state of flood disasters in Japan is described in 4.1.2 (4). Furthermore, repeated drought disasters requiring restrictions on the intake of waterworks water, or the temporary disruption of water supplies to users have recently occurred throughout Japan, and their frequency and duration are both increasing. Regarding the safety of Japan's water supplies by taking the Kiso River System as an example, a flow regime equivalent to water supply stability of 1/10 during the past twenty years has, as shown in Figure 4.2.1, declined from the level when the Water Resources Development Basic Plan was enacted in the 1960s [193]. On the Tone River System, the planned level of water supply stability has been set at 1/5, considering the rising pressure of the water demand in the Tokyo Region, but recently it seems to have fallen to between 1/2 and 1/3 [194].

Future measures against flood disaster will not only include those involving river improvements that benefit a relatively limited range of beneficiaries. Much is also expected of dams that manifest flood control effects in a short time and extend these effects far downstream. Dams must also provide effective drought countermeasures.

Abnormal weather and the roles of dams are being studied by the MLIT, but with new dams not being smoothly constructed, the burden on existing dams will continue to gradually rise. For the time being, the priority issues are redeveloping existing dams and reorganizing groups of existing dams to optimize their operations.

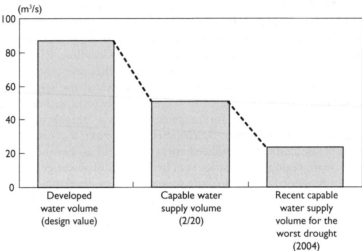

Developed water volume (design value):
　Capable water supply volume, determined on the basis of the river flow regime at the
　time when the Basic Plan for Water Resources Development was drawn up.
Capable water supply volume (2/20):
　Capable water supply volume, determined on the basis of the river flow regime in the
　second worst drought year in the past 20 years (1979–1998).

(Source: The data of National Council Development Subcommittee 2004.5)

Figure 4.2.1 Changes in water supply capacity of the Kiso River System.

Source: MLIT [193].

4.3   FUTURE ROLES OF DAMS

4.3.1   Water environments, water cycles, and dams

As explained in the previous sections, there are factors that reduce and factors that
increase water demand. As a result of the anticipated future population decline in
Japan, the relocation of large offices (universities, hospitals, department stores), and
the qualitative change in agriculture and industry, there are regions where water demand is declining.

This fact signifies a rise in the potential for different water use in such river basins.
In other words, because existing dam reservoirs are major assets, there is leeway for
the redistribution of water supply reservoir capacity, considering the need for water
in a river basin for flood control, water use (including power generation), and water
environments, and the position of effective reservoirs.

In a region where there is insufficient water to conserve a river's downstream environment, for example, there is leeway to reproduce a flow regime matched to a more
natural rhythm by flush-discharging water. In a region with a shortage of agricultural
water, capacity used for other purposes can be switched to agricultural water supply,
and in a region where flood discharge causes frequent damage, the capacity for flood

Figure 4.3.1 Hakusui Dam. (See colour plate section)
Source: Courtesy of the Japan Dam Engineering Center.

control capacity can be increased. Effectively using dams in this way can smoothly and efficiently redistribute water by adapting to challenges in each river basin, and can also greatly contribute to harmonizing human activities with nature in the water environment and water cycle in a river basin.

4.3.2 Reservoir areas and dams

Excluding dams used exclusively for flood control (so-called dry dams), when a dam has been constructed, it forms a reservoir. This reservoir can become an asset for the reservoir region.

Humans have long found serenity at the waterside. So according to geographical conditions such as proximity to a large city or the discovery of hot springs, dam reservoirs have potential capacity of being places for study, recreation, relaxation and festival events: as shoreline resorts, country home sites, water leisure centers, and cherry viewing sites.

Over many years, dam reservoirs create charming water resource regions in natural settings: for example, those at the more than 80 dams that have already been designated as wildlife protection areas.

Many dams such as the Hakusui Dam, which is a historical agricultural facility (Oita Prefecture, Fujioiro Land Improvement District, 1938, Figure 4.3.1) are assets that create a region's scenery and culture in many ways.

4.4 SUSTAINABLE USE OF DAMS SO THEY CAN CONTINUE TO FULFILL THEIR ROLES RELIABLY IN THE FUTURE

4.4.1 Recognition of the roles of dams in basin societies

Even when a dam has been constructed to play an important role anticipated by the basin residents, as time passes after completion, they begin to regard it in a matter-of-fact way, and their consciousness of its existence inevitably fades over time. However, a dam only becomes fully effective after the basin society has recognized its usefulness, so residents must be made aware of the roles of dams.

There are people who expect a great deal of "Green Dams" [195], which means, instead of constructing dams, conserving forests or planting new forests to duplicate the roles of dams. However, the Japan Science Council has stated that, "a forest mitigates flood discharges during medium and small scale floods, but it cannot provide substantial effects during large scale floods". Moreover, it states "Under conditions close to droughty water discharge, there are cases where a forest itself consumes a great deal of water, so it actually lowers the river flow rate" [196].

Consequently, it is important for dam managers to regularly distribute a wide range of information about the roles of dams and strive to provide basin inhabitants with a correct understanding of these roles, so they will realize that dams exist for the benefit of the basin's inhabitants. One way of doing this is to apply the Monitoring and Evaluating System for Dam Management described in the previous chapter.

Based on these experiences, many cases have clearly shown that it is effective and important to create methods of transmitting and providing concerned residents with factual information in an easily understood form about the effectiveness of dams during floods and droughts. Dam operating procedures, particularly during floods, should be efficiently incorporated into regional risk management systems.

4.4.2 Maintaining the functions of dam management facilities, etc.

The most important aspect of dam management is to maintain a dam's functions continuously. To do so, not only must day-to-day dam management, operations and facility inspections be performed properly, but emergency inspections following earthquakes and periodical inspections must also be done.

One important task is the appropriate operation of dams during floods and droughts: which shows how dams can contribute to regional society. An effective way for a dam manager to achieve this goal is to establish a management system that can respond to emergency situations, including those occurring after-hours and during holidays, and ways to link the dam with regional society on a daily basis to permit links with the region during an emergency.

To do so, it is essential that appropriately qualified staff be always assigned and be properly trained.

In addition, in order to ensure that a dam remains effective and fulfills its future roles, its manager has to make repairs following unexpected malfunctioning or damage caused by lightning, earthquakes and so on; replace spillway gates, etc.; make changes to operating systems and information provision facilities in response to technical progress and demands by the public; and carry out needed repairs and improvements based on the results of periodic inspections.

Dam bodies generally have a service life of over a century. Considering the roles that a dam must fulfill, these cannot easily be provided by other means, so it is necessary to renovate gates and other equipment to ensure the dam's required service life.

4.4.3   Sustained use of dam reservoirs

In contrast to dam management equipment that must be replaced over relatively short periods of time, a dam body is presumed to have a service lifetime of at least 100 years, as shown by the fact some dams constructed during the Roman Empire are still in use.

Reservoirs suffer from problems such as sedimentation and water quality. These problems are characteristic of reservoirs and are related to the way that the storage of water in dam reservoirs disrupts the continuity of material cycles. Even if dam bodies are problem-free, the reservoir's functionality is impeded by these problems.

In parts 3.3.2 and 3.3.3, the state of these problems and countermeasures taken in Japan are described.

To achieve the sustainable use of dams, these problems must be resolved by developing more effective technologies.

4.4.4   The effective use of existing dams

Japanese society is changing because of a falling population, and in the future, we must revise the way in which existing water resources are allotted to various uses.

Presented by Mr. Masahisa Okano

Figure 4.4.1 Proserpina Dam. (See colour plate section)

With difficulties facing the construction of new dams, new needs must be satisfied by using existing dams more effectively. Revising the operation (revising the allotment of capacity) of existing dams, revising the operation of multiple dams on a single river system, raising dam height to increase capacity, constructing channels that link reservoirs, and making other facility improvements can further improve dam functions.

To effectively use dams, technological studies of advanced operations are, of course, necessary, and to effectively use existing dams in instances where the dam manager and the organization bearing the cost are different, and where different laws form the basis of their construction and management, management methods, including transfer of rights, compensation, and sharing management costs, and other legal systems, must be provided.

In Japan, such problems are being overcome to use existing dams more effectively; a trend which will presumably accelerate in the future.

4.5 ROLES OF PEOPLE INVOLVED IN DAMS

4.5.1 Development and refinement of dam technologies

Dams made major contributions to the rehabilitation of the national land of Japan following World War II and to economic growth throughout the 20th century by generating electric power, preventing flood disasters, and by supplying water to cities and agricultural users. It was dam engineers who played major roles in creating the fundamental infrastructure of the national land.

Japan's dam engineers have, until now, focused on the construction of safe dams. These dam engineers have been forced to place top priority on guaranteeing the safety of dams, because although dams are now constructed with modern methods, accidents have occurred at dams even in developed countries, and in Japan in particular, there are weather conditions and topographical conditions such as earthquakes and so on that are factors severely impacting the dam construction process.

In the future, dam engineers must respond suitably, not only to ensure the safety of dams, but to take responsibility for sustainable land use and the environment for the benefit of coming generations, the pursuit of economic benefits, and the fulfillment of all their responsibilities to users of public infrastructure (citizens and society), including compliance with the law. They must also strive to develop and refine technologies that will be needed in the future, in line with the engineers' ethical viewpoint to deal with the environmental and social impacts of dam projects.

Technologies needed to manage existing dams must be flexible in order to handle changes in the consciousness of residents on land surrounding dams, changes in management facilities and operating systems (ensuring key personnel), changes in the natural environments of reservoirs and downstream rivers, global warming and other changes in hydrological environments, population decline and other changes in society in reservoir regions, and changes in the effective use of existing dams in response to new needs.

Priority areas where dam engineers will face challenges in the future are not only the development of technologies to construct dam bodies: they will also have to introduce the findings of multi-disciplinary research related to regional societies, ecosystems, the environment, forestry, fisheries, and the leisure industry to establish mitigation, prevention and restoration technologies, and deal with impacts on dams and reservoirs and on the rivers upstream and downstream of their locations. Of course, measures to mitigate social impacts must be taken cooperatively by people in various industries and levels of society, in both the upstream and downstream sections of rivers.

4.5.2 Efforts to conduct linked regional activities

Dam engineers alone cannot complete the roles of dams in regional societies. Dam managers must act as coordinators necessary for cooperation between the many people involved in a regional society, together with various other groups and organizations.

Through such activities, dam engineers must take action to appropriately explain matters related to dams that are of great concern to the public. To mention a few examples: the problem of the impact on the natural environment raised by the World Commission on Dams (WCD) and other groups; the problem of the impact on society caused by the forced relocation of people; the functions of dams and forests; and the environmental problems created by the abandonment of dams.

Dam engineers should, as people involved in dam construction projects and in the management of dams, adequately fulfill their responsibility to provide the residents surrounding dams and citizens of Japan with the information they have obtained, reflecting on their failure in fully providing the information.

4.5.3 International technological cooperation among dam engineers

Many dams in Japan, an earthquake prone country, are as high as dams in other countries, despite their small reservoir capacity.

Japan's dam engineers possess design and construction technologies that they can display proudly to the world, thanks to their past efforts to apply carefully devised technologies that ensure the safety of the dams they construct. Dam engineers include many capable people who have acquired extensive experience in applying diverse technologies, not only in dam body planning, design and execution technologies, but also in areas required to study the regional development effects and economic cost-benefits of flood control, water supply, and electric power generation. They also plan water transportation, water use rights systems and design replacement roads, take action to compensate, relocate, and restore the livelihoods of residents displaced from submerged land, perform surveys and evaluations, take measures to conserve the environment, and to inspect, improve and manage existing dams.

Though the new projects are decreasing in number, a considerable number of high dams are still under construction in Japan (65 dams, as of 2008). And dams are being constructed around the world: in China, India and other countries in Asia, in Africa and Central and South America. The Government of Japan has supported many such dam projects through ODA, etc.

Japan can contribute to resolving international water resource problems by assisting these countries that suffer from severe shortages of food and energy, by applying the knowledge and capabilities listed above and cooperating with their engineers to create technologies suited to the climate, natural features, culture, and society of each country, and by providing technologies to quickly design and construct inexpensive dams with superior durability that can be operated inexpensively and safely.

4.5.4 Preserving technologies for future generations

An important mission of engineers is to ensure that their accumulated technologies and experience are preserved and handed down to future generations.

Humanity has used dam technology to overcome many past crises by guaranteeing drinking water, food, and energy, and by ensuring safety from flooding. Research must be performed to learn how to plan and carry out dam projects to overcome crises whenever they occur or when a new dam becomes necessary. Moreover, dam engineers must respond by passing on this expertise to future generations of engineers.

Specifically, engineers do this in a number of ways: through on-the-job training in practical activities with young engineers, by providing information through mass media, conducting surveys and research through concerned academic associations and technology development organizations, by presenting technical research reports at conferences and publishing technical manuals, and so on.

Engineer Hatta Yoichi – Ushantou Dam [197]

Hatta Yoichi was born in Kanazawa City in Ishikawa Prefecture in 1886. After studying civil engineering, he joined the Government of Formosa in 1910. Around 1917, he took part in the construction of the Ushantou Dam to irrigate the Jianan Plain, a region where farming was impossible because of a shortage of water, and the poorest region in

Presented by Mr. Tatsuo Hamaguchi

Figure 4.5.1 Bronze statue of Hatta Yoichi near the Ushantou Dam.

Taiwan. The work began in 1920 and was completed ten years later in 1930. In 1942, he was ordered to the Philippines, but en route, an American submarine attacked his ship. He went down with the ship at the age of 56.

The irrigation system supplied by the Ushantou Dam that forms a reservoir named the Sanhutan (coral-shaped) Lake is called the Jianan Canal. It irrigates 150,000 hectares and benefits 150,000 farming households. The Jianan Canal is operated smoothly in conjunction with a water distribution method that uses the three-year cycle proposed by Hatta Yoichi, thereby transforming the Jianan Plain into rich farmland and bringing astonishing improvements to the daily life of its people. The local farmers have not forgotten Hatta Yoichi, and continue to preserve a bronze statue erected in his honor (Figure 4.5.1).

Many Japanese caught up in the present wave of internationalization now work in countries around the world. Their most important activity is probably the pursuit of the well-being of the local people. The situation in Taiwan, which was under Japanese rule at that time, differs from that under today's international cooperation, but the achievements of Hatta Yoichi who devoted himself to local development continues to provide benefits today. This is the great dream of civil engineering and is linked to our ways of living in the international society of the future.

Chapter 5

Summation

It is often said that water is an integral part of civilization. History has confirmed that many of the world's civilizations collapsed due to the disappearance of their water.

The importance of water also applies to Japan, which is considered to have abundant water resources. Chapter 2 explains the roles that dams have played in Japan as mechanisms of using and controlling water, and gives the background to each stage in their history. This approach has shown that dams have played important roles at each turning point in Japan's history. It also enabled us to discover the hard work done by previous generations to secure the water they needed. This book focuses on dams, but other facilities were also constructed and water management was performed, so the cases described in this book are part of the overall picture, and many other measures and cases have been omitted.

Chapter 3 describes environmental and social impacts of dams in Japan, discussing the negative impacts of dams and countermeasures taken. It is true that people have recently become negative toward dams in light of their great impact on society, the environment and the economy. This is true not only in Japan, but also worldwide, as shown by the WCD Report (Chapter 1), and the 3rd Word Water Forum Reports, etc. The WCD Report points out that dams place an excessive burden on residents, that they severely impact the environment, and that they do not provide functionality matched to the investment required to construct them. Case studies have revealed the problems caused by dams. Many measures have already been taken in Japan to deal with the problems identified, but efforts must continue to avoid or reduce their impact. Nevertheless, conditions surrounding dams differ greatly from country to country, as clearly shown by these reports and by the 3rd World Water Forum (Kyoto 2003, with 24,000 delegates from 150 countries). The key is to clarify the positive and negative impacts of dams on local societies and to find ways to balance these impacts.

Chapter 4 also touches on this, but even in Japan, where some argue that dams are unnecessary, almost every year, floods and droughts cause disasters throughout the country. It has been pointed out that abnormal climatic conditions have caused imbalanced rainfall and have gradually increased their intensity. We now face many water problems, including water management in river basins, environmental deterioration and global environmental problems. It would be an oversimplification to claim that these problems can be resolved by not constructing dams. Today, as water shortages worsen, instead of simply ignoring these problems, we must strive harder to resolve them.

This work has described the roles of dams in Japan, but it is impossible to discuss the history of dam construction as a single process because it differs greatly, according to national conditions, the historical period and the region. However, the history of dam construction has four distinct stages: the stage where dams were constructed to serve small districts; the stage where urbanization required dams to be constructed systematically to serve a large number of people; the stage where problems that dams cause to both society and the environment were clearly revealed; and the stage where water problems have been reduced and residents' consciousness of dams has gradually weakened, resulting in an increase in the number of people who argue that dams are unnecessary, or that they should be eliminated. Japan is not an exception to this process. This report has explained the measures that Japan has taken in response to each of the above stages.

We encourage everyone working in the field of dams to refer to this report as they compare cases in various countries and dams that they have worked on.

Outline of the National Administrative Organs of Japan

1.1 OUTLINE

The governing system of Japan is divided into judicial, legislative and administrative branches. The judiciary has a three-level system consisting of the Supreme Court, higher courts and regional courts. The legislature is a two-house system consisting of the House of Representatives and the House of Councilors. The Administration consists of one office and twelve ministries led by the Prime Minister.

Regional administration has a two-tier system: prefectures (a total of 47) and cities, towns and villages (approximately 1,800 in 2007). Local autonomy by local government is stipulated by the constitution, and each has a unicameral assembly and an administrative organization.

The present structure of the central government has been formed by reorganizing its ministries and agencies, which was done in January 2001. This reorganization was a sweeping consolidation of the previous single office and 22 ministries (Table A1.1).

Table A1.1 National Administrative Organs.

Cabinet Office
National Public Safety Commission (National Police Agency)
Ministry of Internal Affairs and Communications
Ministry of Justice
Ministry of Foreign Affairs
Ministry of Finance
Ministry of Education, Culture, Sports, Science and Technology
Ministry of Health, Labour and Welfare
Ministry of Agriculture, Forestry and Fisheries (MAFF)
Ministry of Economy, Trade and Industry (METI)
Ministry of Land, Infrastructure, Transport and Tourism (MLIT)
Ministry of the Environment (MOE)
Ministry of Defense

2.1   ADMINISTRATIVE ORGANIZATION RELATE TO RIVERS, DAMS, WATER RESOURCES, ETC.

River management, including flood control, falls under the jurisdiction of the MLIT. This ministry was formed by combining the Ministry of Construction, Ministry of Transport, National Land Agency and the Hokkaido Development Bureau. Prior to the reorganization, river management was under the jurisdiction of the former Ministry of Construction. The MLIT constructs flood control and multi-purpose dams and authorizes water rights and the construction of water use dams. It also subsidizes river management carried out by local governments.

Matters concerning sewage systems (urban stormwater discharge and wastewater disposal) also fall under the jurisdiction of the MLIT. Sewage systems are constructed and operated by local governments, with the MLIT supervising and subsidizing their construction and operation.

Matters concerning agriculture fall under the jurisdiction of the MAFF. This ministry constructs irrigation dams and prepares agricultural land. It also supervises and subsidizes agricultural projects implemented by local governments.

Matters concerning waterworks fall under the jurisdiction of the Ministry of Health, Labour and Welfare. Local governments ensure sources for waterworks and construct and operate waterworks facilities, while the Ministry of Health, Labour and Welfare supervises and subsidizes these tasks. The Ministry of Health, Labour and Welfare is an organization that was formed by combining the Ministries of Welfare and Labour. Prior to reorganization of the agencies and ministries, the Ministry of Welfare administered waterworks.

Matters concerning electric power generation and industrial water fall under the jurisdiction of the METI (prior to the reorganization of ministries and agencies, the Ministry of International Trade and Industry, MITI). Electric power companies construct and operate electric power generation facilities while local governments construct and operate industrial water supply systems, with the METI supervising and subsidizing these tasks.

Water quality standards for rivers, lakes, and marshes, environmental assessments and other matters related to the environment fall under the jurisdiction of the MOE (the Environmental Agency prior to the reorganization of the agencies and ministries). Actual hands-on environmental administration in each district is performed by local government, with the MOE supervising and subsidizing their environmental administration.

The basic framework of river management system in Japan

I OUTLINE OF THE RIVER LAW

The foundational law for river management in Japan is the River Law, which is administered by the MLIT. The present River Law was enacted in 1964 then revised in 1997 (to add to its purposes the improvement and conservation of river environments).

Its predecessor, the Previous River Law that was enacted in 1896 (administered by the former Ministry of the Interior), prioritized flood disaster prevention, stipulating that in principle, rivers were managed by prefectures, but that, as necessary, the national government performed management and implemented projects. In 1935, the former Ministry of the Interior enacted the River Dam Regulation that prescribed the authorization of the construction of water use dams. River improvement projects under the Previous River Law were carried out with great success throughout Japan, but it did not provide fully for the integrated use of river water, so in 1964, it was completely revised to create the present River Law. At that time, the Ministry of Construction administered the River Law, but a reorganization of ministries and agencies carried out in January 2001 integrated the Ministry of Transport, the Ministry of Construction and other agencies, creating the MLIT that now administers the River Law.

In Japan, legislation is organized in a hierarchical structure with the Constitution at the top, followed by Laws, Cabinet Orders, and Cabinet Office and Ministry Ordinances. A Law is enacted by a decision of the Diet. A Cabinet Order that stipulates a regulation necessary to enforce a Law and matters delegated by a Law are enacted by the Cabinet. Cabinet Office and Ministry Ordinances are enacted by the Prime Minister, or by another Cabinet Minister, concerning administrative matters under their authority. Regarding the River Law, related Cabinet Orders include the River Law Enforcement Order and the Cabinet Order Concerning Structural Standards for River Management Facilities, etc. Related Ministry Ordinances include the Ordinance for the Enforcement of the River Law. Government Directives and Notifications are issued to establish guidelines for more specific administrative operations. Government Directives are issued to organs under each Minister's authority and to their employees. Notifications are issued by an administrative organ to other organs under its jurisdiction and to their employees, and they include Administrative Vice-Ministers' Notifications, Deputy Director Generals' Notifications, and Section Directors' Notifications.

The present River Law stipulates, in Article 1, the purpose of the Law as follows: "The purpose of the Law is to contribute to the conservation and the development of national land, and thereby sustain the safety of the public and promote public welfare, by performing integrated control of rivers to prevent the occurrence of disasters caused by flooding or storm surges, utilizing rivers appropriately, maintaining the normal functions of river water and improving and conserving river environments." In other words, its purposes are integrated management: controlling flood discharge, water usage and the environment. As the general principle governing river management, it stipulates that the river is a public property and that the river water cannot be the object of private rights (Article 2).

A major characteristic of the present River Law is Integrated River System Management. This categorizes river systems in light of national land conservation and the national economy. Particularly important river systems are defined as Class A Rivers and are managed centrally by the national government. Other rivers with important links with the public interest are defined as Class B Rivers and they are each managed by a prefecture. Class A Rivers are managed by the Minister of Land, Infrastructure, Transport and Tourism, and Class B Rivers are managed by prefectural governors.

The Law clearly defines dams installed on rivers as, "a structure constructed with the authorization of the river manager to store or to take water flowing in a river, and that has a height of 15 m or more from the foundation ground to the crest of the dam" (Article 44, paragraph 1). Constructing a dam on a river is generally: (1) exclusively occupying land along a river; (2) excavating the ground in the river; (3) installing a structure (dam); and (4) exclusively using the river water, and authorization for each of these actions must be given by the river manager (Article 24, Article 27, Article 26 and Article 23, respectively).

Special Provisions for Dams comprise the following:

> Maintaining a river's former functions (Article 44), observing the water level and flow rate (Article 45), reporting on the state of operation of a dam (Article 46), dam operation regulations (Article 47), measures to prevent danger (Article 48), preparation of records, etc. (Article 49), assigning engineers in charge of management (Article 50) and exceptional provisions for dams that are joint use structures (Article 51).

In addition to the River Law, which is the foundation of river management, other related legal systems have been established: the Specified Multi-purpose Dam Law, the Water Resources Development Promotion Law, the Law Concerning Special Measures for Reservoir Area Development and the Sabo (Erosion Control) Law that concerns the prevention of sediment disasters.

2   RIVER MANAGERS: ROLES AND ORGANIZATIONS

There are 109 river systems throughout Japan that have been designated as Class A Rivers. The total area of their river basins is 246,900 km², or 63% of all the land of Japan. (The average river basin area per river system is about 2,200 km².) The principal sources of water used by the large metropolitan regions, Tokyo, Osaka, and Nagoya are Class A River Systems: the Tone River, Yodo River and the Kiso River, respectively. There are 2,723 river systems that are designated as Class B Rivers in Japan. The total area of their river basins is 109,400 km². (The average river basin area per river system is about 40 km².) (Note: values of Class A and B Rivers are based on the 2006 River Annual.)

As stated earlier, the river manager of Class A Rivers is the national government (the Minister of Land, Infrastructure, Transport and Tourism), and the river managers of Class B Rivers are prefectural governors, but a study of the details shows that relationships between the national government and local governments are a little more complex. On a Class A River, for example, it is difficult for the national government to manage the entire river, so prefectural governors are obligated to manage parts of the rivers that are less important. This means that important and other parts of Class A Rivers are managed by the national government and by prefectures, respectively. The River Law also stipulates that even on Class B Rivers, prefectural governors should obtain agreement from the Minister of Land, Infrastructure, Transport and Tourism for action taken to deal with electric power generation and river water usage, in excess of a specified scale (River Law Article 79 paragraph 2). Part of the work performed to improve river environments may be delegated to the municipalities.

The roles of a river manager are broadly categorized into two areas: constructing and managing river management facilities including dams, levees, drainage equipment and issuing authorizations concerning rivers.

Firstly, a river manager provides flood protection, maintains the normal functions of river water, and improves and conserves the river environment. Examples of this work are constructing levees, dredging river courses and constructing and managing dams to control flooding. To carry out these river improvements, they enact a Fundamental River Management Policy and prepare a River Improvement Plan, based on the River Law. On Class B Rivers, projects are led by prefectures, but large-scale projects are generally undertaken with subsidization by the national government, and in such cases, the national government provides guidance for the projects.

River managers issue authorizations under the River Law when any party other than the river manager wishes to carry out any activity in a river area. Examples of actions requiring authorization are bridging a river, constructing a dam on a river, or taking water from a river. Even when a project is initiated by the national government or a prefecture, they must obtain authorization from, or consult with, the river manager.

River management work on behalf of the MLIT is performed by the River Bureau (with approximately 260 employees, more than half of which were engineers, in 2007). Regional branches of the River Bureau are the river departments in the Regional Bureaus (a total of eight nationally), the Hokkaido Development Bureau, and the Okinawa General Office of the Cabinet Office. Between 20 and 30 offices that are operated under each Regional Bureau execute projects, handle authorizations and manage dams. In addition, the Japan Water Agency (JWA, reorganized from Water Resources Development Corporation in 2003) develops and manages water resources on specified rivers, including the Tone, Ara and Yodo Rivers.

Technological development plays an important role in the appropriate implementation of projects and in issuing authorizations. Therefore, the river divisions of the National Institute for Land and Infrastructure Management and the Public Works Research Institute, both part of the MLIT, develop new technologies and resolve technological problems in the field. Although not administrative bodies of the national government, the Japan Dam Engineering Center, the Water Resources Environment

Technology Center and other non-profit corporations support river management work of the national government and local governments and develop new technologies under the supervision of the responsible ministry.

In prefectural administrations, river departments within public works bureaus manage rivers. Like the MLIT, their headquarter organizations are supported by work conducted in the field by public works offices.

The national government's budget system provides for the cost of river management from its Special Account for River Management. This includes the cost of constructing dams and levees, implementing other flood control projects, maintaining related facilities and the personnel cost of national government employees who are involved in river management work. The Special Account for River Management in fiscal 2006 was approx. 1.18 trillion yen ($9.8 billion: calculated at $1.00 = 120 yen), with approx. 30 billion yen ($2.5 billion) earmarked for dam projects. Approximately 8,160 employees were paid from this account, and more than half of these were engineers (in-house engineers).

3 OUTLINE OF THE LEGAL FRAMEWORK CONCERNING CONSTRUCTION AND SAFETY INSPECTIONS OF WATER USE DAMS

Figure A2.1 shows a typical example of the legal procedures followed when constructing a hydropower or irrigation dam (so-called water use dams) with a dam height in excess of 15 m, on a Class A River.

A structure with a height lower than 15 m is handled as a "weir". This too requires authorization under the River Law, but the technological standards applied to it differ from those applied to dams.

As for structural safety, Article 13 of the River Law stipulates the basic concept that the technological standards be determined in terms of a Cabinet Order. Namely, Article 13 paragraph 1, states, "River management facilities or a structure installed with authorization pursuant to Article 26 paragraph 1 shall be structurally safe, considering the water level, flow rate, topography, geology and other characteristics of the river, and considering self weight, water pressure and other predicted loads." Paragraph 2 states, "Technological standards considered essential to river management for dams, levees, or other important structures shall be stipulated by the Cabinet Order."

In 1976, the Cabinet Order Concerning Structural Standards for River Management Facilities was enacted, based on these provisions. Chapter II of the Cabinet Order Concerning Structural Standards (Article 4 through Article 16) contains provisions concerning dams, stipulating the principles of the structures (Article 4), height of the non-overflow part of a dam (Article 5), types of loads acting on a dam (Article 6), installation of spillways (Article 7) and other technological standards. The Enforcement Regulation of the Cabinet Order Concerning Structural Standards for River Management Facilities, which is a Ministry Ordinance, provides even more specific regulations based on these provisions.

Article 30 of the River Law stipulates that, "The river manager shall carry out a completion inspection of the dam, and until it passes this inspection, the dam shall not be used."

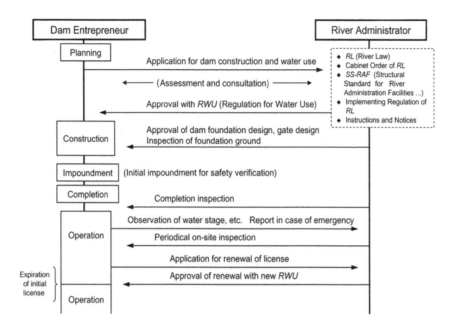

Figure A2.1 Typical example of legal procedures for a water use dam.

Prior to the completion inspection, it is important to perform initial impoundment in order to confirm overall safety by verifying structural safety and impermeability of the dam body, as well as by checking the slope of the reservoir for landslides. This initial impoundment is performed by, in principle, gradually raising the reservoir water level to the surcharge water level while performing a variety of measurements and observations, maintaining this water level for a specified period of time, and then gradually lowering the water level again. This is usually done for a period of six months to one year. If a problem is noticed during this period, the necessary measures are taken to resolve it.

Meticulous observation that is performed during the initial impoundment after a dam body has been completed to confirm its safety is done in many countries, but performing initial impoundment to confirm safety in such a thorough fashion is a feature unique to dam construction in Japan. Initial impoundment for safety verification has been done because of past problems with frequent landslides having occurred along reservoirs. Doing this, primarily to confirm the effectiveness of landslide prevention work being done as part of dam projects, began with dams implemented directly by the MLIT in the late 1970s. In the 1980s, the execution of this step became a fixed rule for projects supervised by the MLIT. At dams for hydropower and other water use, its performance began after initial impoundment for safety verification at MLIT dam projects became the norm. Although initial impoundment for safety verification is not required by law, it is now done without exception under the guidance of river managers.

Appendix 3

Outline of the existing dams in Japan

At least from 7th century, Japanese people have been constructing dams to make their life better. In the background of this restless efforts lie the following characteristics of Japan relevant to the topographic and climatic conditions [198].

♦ Japan consists of a chain of narrow islands with mountain ranges at their backbone, which form many small river basins with steep streams.
♦ Around 2/3 of the national land is covered with forests, rich in environmental assets.
♦ Japan belongs to the circum-Pacific earthquake belt, and is subject to strong earthquake disasters quite frequently.
♦ The geological constitution is very complicated, and sediment production in mountains is high.
♦ Major floods are brought by the "Tsuyu" rainy season in June and typhoons in autumn, whose rainfall is characterized by high intensity and short duration. Most floods end within 2 or 3 days. Snow-melt floods are often seen in the northern part of Japan.
♦ Alluvial plains spread in the lower reaches. Population and economic activities concentrated there which need to be protected from flood disasters. This is why Japan has a long history of people's efforts to overcome these disasters.

Table A3.1 shows the major 10 river systems in Japan in terms of the basin area. Even the largest one, the Tone River System, has the basin area of 16,840 km^2, which is quite small when compared with those in continents. It is also understood that the scale of the basin is very compact judging from the short river length.

The Tone, Kiso and Yodo River Systems are important supply source of water for the Tokyo, Nagoya and Kyoto-Osaka-Kobe metropolitan area respectively.

In correspondent with these features, dams in Japan have relatively small reservoirs. In view of water resources development by dam reservoirs, water supply is usually achieved by combining run-of-the-river flow and replenishment from the reservoir, which enables efficient supply of water.

As for flood control operation of dams, small catchment area and torrential rainfalls demand accurate operation in a limited time. In case the catchment area is very small, flood control dams are often planned as non-gate dams, not as gated ones.

Table A3.1 Major river systems in Japan.

Ranking	Name of the river system	Region	Basin area (km²)[1]	Length of trunk river (km)[2]	Population in the basin (million)[3]
1	Tone	Kanto	16,840	322	12.65
2	Ishikari	Hokkaido	14,330	268	2.59
3	Shinano	Hokuriku	11,900	367	2.97
4	Kitakami	Tohoku	10,150	249	1.39
5	Kiso	Chubu	9,100	229	1.93
6	Tokachi	Hokkaido	9,010	156	0.34
7	Yodo	Kansai	8,240	75	10.92
8	Agano	Hokuriku	7,710	210	0.58
9	Mogami	Tohoku	7,040	229	1.00
10	Teshio	Hokkaido	5,590	256	0.09

1, 2 Based on the web site the Statistics Bureau, Ministry of Internal Affairs and Communication (http://www.stat.go.jp/data/chouki/01.htm).
3 Based on the data of the River Bureau, MLIT (http://www.mlit.go.jp/river/toukei_chousa/kasen/ryuiki.pdf).

Table A3.2 Number of dams by year of completion.

Year of completion	Dam height		Total	Share (%)
	Over 30 m	15–29 m		
Before 1000 AD	1	8	9	0.3
1001–1899	3	658	661	21.6
1900–1945	85	607	692	22.6
1946–1999	884	650	1,534	50.2
2000–2008	132	30	162	5.3
Total	1,105	1,953	3,058	100

Based on the data of JCOLD as of March 2008: Large dams, No. 204 p. 100, 2008.

The number of so-called large dams with height of 15 m or more is shown in Table A3.2. As of March 2007, there are as many as 3,058 existing large dams in Japan, more than half of which were completed during about 50 years' period after the World War II.

Table A3.3 shows the number of existing dams counted by purpose. Dams whose project purpose includes irrigation occupy 52.7%, followed by dams with flood control purpose (16.7%) and dams with hydropower purpose (16.2%).

Table A3.4 shows major 30 dams with large reservoir capacity. Tokuyama Dam, a multi-purpose dam completed by JWA in 2008, has the largest capacity of 660 million m³. This capacity may look small when compared with multi-billion cubic meter reservoirs in US or Europe; however, its regulating capability to the river flow is quite substantial in view of the small basin scale.

Tables A3.5–A3.11 shows major high dams by dam type. The highest dam in Japan is the Kurobe Dam, an arch type hydropower dam on the Kurobe River. These table will provide general idea of existing high dams in Japan.

Table A3.3 Number of existing dams by purpose.

Purpose	Dam height		Total	Share (%)
	Over 30 m	15–29 m		
Irrigation	460	1588	2048	52.7
Waterworks/ Industrial water	365	136	501	12.9
Hydropower	433	197	630	16.2
Flood control	532	118	650	16.7
Others	53	5	58	1.5
Total	1843	2044	3887	100
(Multi-purpose dam)	(458)	(86)	(544)	

Based on the data of JCOLD as of March 2008: Large dams, No. 204 p. 100, 2008.

Remarks: Total number is not equal to that of Table 2, since the number is counted twice of more, if the dam is of multi-purpose.

Table A3.4 Major dams with large reservoir capacity.

Ranking	Name of dam	River system	Prefecture	Dam type	Purpose	Dam height (m)	Dam length (m)	Dam body volume (1000 m³)	Reservoir area (ha)	Storage capacity (million m³)	Catchment area (Km²)	Leading owner	Fiscal year of completion	Ref. no.*
1	Tokuyama	Kiso	Gifu	ER	CNSH	161	427	13,700	1,300	660	255	JWA	2007	75
2	Okutadami	Agano	Niigata, Fukushima	PG	H	157	480	1,636	1,150	601	426	J-POWER	1960	120
3	Tagokura	Agano	Fukushima	PG	H	145	462	1,950	995	494	702	J-POWER	1959	
4	Miboro	Sho	Gifu	ER	H	131	405	7,950	880	370	14	J-POWER	1961	120
5	Kuzuryu	Kuzuryu	Fukui	ER	CH	128	355	6,300	890	353	185	MLIT	1968	
6	Ikehara	Shingu	Nara	VA	H	111	460	647	843	338	277	J-POWER	1964	
7	Sakuma	Tenryu	Shizuoka	PG	H	156	294	1,120	715	327	4157	J-POWER	1956	82
8	Sameura	Yoshino	Kochi	PG	CNISH	106	400	1,187	750	316	527	JWA	1978	
9	Hitotsuse	Hitotsuse	Miyazaki	VA	H	130	416	555	686	261	415	Kyushu EPCO	1963	117
10	Tamagawa	Omono	Akita	PG	CNISH	100	442	1,150	830	254	287	MLIT	1990	
11	Uryu No. 1	Ishikari	Hokkaido	PG	H	46	216	188	2,373	245	203	HEPCO	1943	
12	Tedorigawa	Tedori	Ishikawa	ER	CSH	153	420	10,050	525	231	428	MLIT	1979	60
13	Takami	Shizunai	Hokkaido	ER	CH	120	435	5,120	675	229	283	HEPCO	1983	
14	Arimine	Joganji	Toyama	PG	H	140	500	1,568	512	222	50	Hokuriku EPCO	1959	

No.	Name	River	Prefecture									Owner	Year	Ref.
15	Yagisawa	Tone	Gumma	VA	CNISH	131	352	510	570	204	167	JWA	1967	31
16	Kurobe	Kurobe	Toyama	VA	H	186	492	1,582	349	199	185	KEPCO	1963	58
17	Nukabira	Tokachi	Hokkaido	PG	H	76	293	460	822	194	388	J-POWER	1956	
18	Miyagase	Sagami	Kanagawa	PG	CNSH	156	375	2,000	460	193	214	MLIT	2001	52
19	Ogochi	Tama	Tokyo	PG	SH	149	353	1,676	425	189	263	TEPCO	1957	49
20	Iwaya	Kiso	Gifu	ER	CISH	128	366	5,780	426	174	150	JWA	1976	
21	Abugawa	Abu	Yamaguchi	PG/VA	CNH	95	286	427	420	154	523	Pref. Gov.	1974	
22	Surikami-gawa	Abukuma	Fukushima	ER	CNISH	105	719	8,300	460	153	160	MLIT	2006	
23	Kanayama	Ishikari	Hokkaido	HG	CNSH	57	289	220	920	150	470	MLIT	1967	77
24	Ikawa	Oi	Shizuoka	HG	H	104	243	430	422	150	459	CEPCO	1957	15
25	Tase	Kitakami	Iwate	PG	CIH	82	320	420	600	147	740	MLIT	1954	
26	Niikappu	Niikappu	Hokkaido	ER	H	103	326	3,071	435	145	193	HEPCO	1974	40
27	Shimokubo	Tone	Gumma	PG	CNSH	129	605	1,193	327	130	323	JWA	1968	
28	Kazeya	Shingu	Nara	PG	H	101	330	592	446	130	660	J-POWER	1960	
29	Shin–Takahashi	Takahashi	Okayama	PG	SH	103	289	430	360	128	635	ENERGIA	1968	102
30	Nariwagawa Oku–Miomote	Miomote	Niigata	VA	CNH	116	244	257	430	126	175	Pref. Gov.	2001	

* Reference number corresponds with the number in the table "Main Dimensions of Dams"
Based on the data of Dam Almanac 2007, Japan Dam Foundation.

Table A3.5 Major high concrete gravity dams (PG).

Ranking	Name of dam	River sytem	Prefecture	Dam type	Purpose	Dam height (m)	Dam length (m)	Dam body volume (1000 m³)	Reservoir area (ha)	Storage capacity (million m³)	Catchment area (Km²)	Leading owner	Fiscal year of completion	Ref. no.*
1	Okutadami	Agano	Niigata/Fukushima	PG	H	157	480	1,636	1,150	601	426	J-POWER	1960	
2	Urayama	Ara	Saitama	PG	CNSH	156	372	1,750	120	58	52	JWA	1999	43
3	Miyagase	Sagami	Kanagawa	PG	CNSH	156	375	2,000	460	193	214	MLIT	2001	52
4	Sakuma	Tenryu	Shizuoka	PG	H	156	294	1,120	715	327	4,157	J-POWER	1956	82
5	Ogochi	Tama	Tokyo	PG	SH	149	353	1,676	425	189	263	Pref. Gov	1957	49
6	Tagokura	Agano	Fukushima	PG	H	145	462	1,950	995	494	702	J-POWER	1959	
7	Arimine	Joganji	Toyama	PG	H	140	500	1,568	512	222	50	Hokuriku EPCO	1959	
8	Kusaki	Tone	Gumma	PG	CNISH	140	405	1,321	170	61	254	JWA	1979	35
9	Takizawa	Ara	Saitama	PG	CNSH	140	424	1,800	145	63	109	JWA	2007	41
10	Shimokubo	Tone	Gumma/Saitama	PG	CNSH	129	605	1,193	327	130	323	JWA	1968	40

* Reference number corresponds with the number in the table "Main Dimensions of Dams"
Based on the data of Dam Almanac 2007, Japan Dam Foundation.

Table A3.6 Major high arch dams (VA).

Ranking	Name of dam	River sytem	Prefecture	Dam type	Purpose	Dam height (m)	Dam length (m)	Dam body volume (1000 m³)	Reservoir area (ha)	Storage capacity (million m³)	Catchment area (Km²)	Leading owner	Fiscal year of completion	Ref no.*
1	Kurobe	Kurobe	Toyama	VA	H	186	492	1,582	349	199	185	KEPCO	1963	58
2	Nukui	Ota	Hiroshima	VA	CNSH	156	382	810	160	82	253	MLIT	2001	
3	Nagawado	Shinano	Nagano	VA	H	155	356	660	274	123	381	TEPCO	1969	
4	Kawaji	Tone	Tochigi	VA	CNIS	140	320	700	220	83	324	MLIT	1983	24
5	Takane No. 1	Kiso	Gifu	VA	H	133	276	330	117	44	125	CEPCO	1969	
6	Yagisawa	Tone	Gumma	VA	CNISH	131	352	510	570	204	167	JWA	1967	31
7	Hitotsuse	Hitotsuse	Miyazaki	VA	H	130	416	555	686	261	415	Kyushu EPCO	1963	
8	Managawa	Kuzuryu	Fukui	VA	CNH	128	357	507	293	115	224	MLIT	1977	61
9	Kawamata	Tone	Tochigi	VA	CNH	117	131	147	259	88	324	MLIT	1966	27
10	Shin–Toyone	Tenryu	Aichi	VA	H	117	311	348	156	54	136	J-POWER	1972	

* Reference Number corresponds with the number in the table "Main Dimensions of Dams"
Based on the data of Dam Almanac 2007, Japan Dam Foundation.

Table A3.7 Major high concrete hollow gravity dams (HG).

Ranking	Name of dam	River system	Prefecture	Dam type	Purpose	Dam height (m)	Dam length (m)	Dam body volume (1000 m³)	Reservoir area (ha)	Storage capacity (million m³)	Catch-ment area (Km²)	Leading owner	Fiscal year of completion	Ref. no.*
1	Hatanagi No. 1	Oi	Shizuoka	HG	H	125	292	598	251	107	318	CEPCO	1962	77
2	Ikawa	Oi	Shizuoka	HG	H	104	243	430	422	150	459	CEPCO	1957	
3	Uchinokura	Kaji	Niigata	HG	CISH	83	166	216	100	25	48	MAFF	1972	
4	Yokoyama	Kiso	Gifu	HG	CIH	81	220	320	170	43	471	MLIT	1964	
5	Ohmorigawa	Yoshino	Kochi	HG	H	73	191	146	92	19	22	YONDEN	1959	
6	Hatanagi No. 2	Oi	Shizuoka	HG	H	69	171	155	49	11	329	CEPCO	1961	
7	Takane No. 2	Kiso	Gifu	HG	H	69	232	162	58	12	173	CEPCO	1968	
8	Ananaigawa	Yoshino	Kochi	HG	H	67	252	219	195	46	53	YONDEN	1963	
9	Zao	Mogami	Yamagata	HG	CNS	66	274	276	24	7	21	Pref. Gov.	1970	
10	Kawamoto	Takahashi	Okayama	HG	CH	60	259	216	80	17	226	Pref. Gov.	1964	

* Reference number corresponds with the number in the table "Main Dimensions of Dams"
Based on the data of Dam Almanac 2007, Japan Dam Foundation.

Table A3.8 Major high buttress dams (CB).

Ranking	Name of dam	River system	Prefecture	Dam type	Purpose	Dam height (m)	Dam length (m)	Dam body volume (1000 m³)	Reservoir area (ha)	Storage capacity (million m³)	Catchment area (Km²)	Leading owner	Fiscal year of completion	Ref. no.*
1	Marunuma	Tone	Gumma	CB	H	32	88	14	68	14	21	TEPCO	1931	
2	Sasanagare	Kameda	Hokkaido	CB	S	25	199	36	7	0.6	–	Hakodate City	1923	8
3	Ombara	Yoshii	Okayama	CB	H	24	94	30	26	2	24	ENERGIA	1928	
4	Mitaki	Chiyo	Tottori	CB	H	24	83	9	3	0.2	22	ENERGIA	1937	
5	Mattate	Joganji	Toyama	CB	H	22	61	4	1	0.03	–	Hokuriku EPCO	1929	

* Reference number corresponds with the number in the table "Main Dimensions of Dams"
Based on the data of Dam Almanac 2007, Japan Dam Foundation.

Table A3.9 Major high rockfill dams (ER).

Ranking	Name of dam	River system	Prefecture	Dam type	Purpose	Dam height (m)	Dam length (m)	Dam body volume (1000 m³)	Reservoir area (ha)	Storage capacity (million m³)	Catchment area (Km²)	Leading owner	Fiscal year of completion	Ref. no.*
1	Takase	Shinano	Nagano	ER	H	176	362	11,590	178	76	131	TEPCO	1979	63
2	Tokuyama	Kiso	Gifu	ER	CNSH	161	427	13,700	1,300	660	255	JWA	2007	75
3	Naramata	Tone	Gumma	ER	CNISH	158	520	13,100	200	90	60	JWA	1990	32
4	Tedorigawa	Tedori	Ishikawa	ER	CSH	153	420	10,050	525	231	428	MLIT	1979	60
5	Misogawa	Kiso	Nagano	ER	CNISH	140	447	8,900	135	61	55	JWA	1996	
6	Minamiaiki	Shinano	Nagano	ER	H	136	444	7,300	59	19	6	TEPCO	2005	
7	Miboro	Sho	Gifu	ER	H	131	405	7,950	880	370	14	J-POWER	1961	120
8	Kuzuryu	Kuzuryu	Fukui	ER	CH	128	355	6,300	890	353	185	MLIT	O1968	
9	Iwaya	Kiso	Gifu	ER	CNISH	128	366	5,780	426	174	1,035	JWA	1976	
10	Nanakura	Shinano	Nagano	ER	H	125	340	7,380	72	33	150	TEPCO	1978	

* Reference Number corresponds with the number in the table "Main Dimensions of Dams"
Based on the data of Dam Almanac 2007, Japan Dam Foundation.

Table A3.10 Major high earthfill dams (TE).

Ranking	Name of dam	River system	Prefecture	Dam type	Purpose	Dam height (m)	Dam length (m)	Dam body volume (1000 m³)	Reservoir Area (ha)	Storage capacity (million m³)	Catchment area (Km²)	Leading owner	Fiscal year of completion	Ref no.*
1	Seiganji	kuma	Kumamoto	TE	CI	61	199	586	19	3	18	Pref. Gov.	1978	
2	Okuboyama	Sozu	Ehime	TE	IS	56	170	491	6	0.8	6	Pref. Gov.	1978	
3	Fukada–Choseichi	Abukuma	Fukushima	TE	I	56	340	1,179	50	9	1	MAFF	1982	
4	Nagara	Murata	Chiba	TE	S	52	250	1,455	81	10	3	JWA	1993	
5	Tateiwa	Tateiwa	Ehime	TE	I	48	175	367	6	0.8	7	Pref. Gov.	1980	
6	Choshi	Shigenobu	Ehime	TE	I	47	142	311	7	0.8	4	Pref. Gov.	1977	
7	Asakura	Tonden	Ehime	TE	I	47	253	531	9	1.4	8	Pref. Gov.	1981	
8	Nakazato	Inabe	Mie	TE	IS	46	985	2,970	130	16	4	JWA	1976	
9	Sawada	Ishida	Niigata	TE	I	46	129	931	14	2	1	Pref. Gov.	1988	
10	Oura	Takori	Saga	TE	I	45	160	355	7	0.7	2	Pref. Gov.	1987	

* Reference Number corresponds with the number in the table "Main Dimensions of Dams"
Based on the data of Dam Almanac 2007, Japan Dam Foundation.

Table A3.11 Major high asphalt facing dams (FA).

Ranking	Name of dam	River system	Prefecture	Dam type	Purpose	Dam height (m)	Dam length (m)	Dam body volume (1000 m³)	Reservoir area (ha)	Storage capacity (million m³)	Catchment area (Km²)	Leading owner	Fiscal year of completion	Ref. no.*
1	Yashio	Naka	Tochigi	FA	H	91	263	2,109	47	12	2	TEPCO	1995	
2	Miyama	Naka	Tochigi	FA	ISH	76	334	1,967	97	26	53	MAFF	1973	
3	Oseuchi	Omaru	Miyazaki	FA	H	66	166	860	27	6	2	Kyushu EPCO	2006	
4	Tataragi	Maru yama	Hyogo	FA	H	65	278	1,462	105	19	13	KEPCO	1974	
5	Futaba	Shiri betsu	Hokkaido	FA	I	61	248	660	64	10	63	MLIT	1987	
6	Otsumata	Agano	Fukushima	FA	H	52	163	320	10	2	180	J-POWER	1968	
7	Kanasumi	Omaru	Miyazaki	FA	H	43	140	390	27	–	2	Kyushu EPCO	2006	
8	Numap-para	Naka	Tochigi	FA	H	38	1597	1,260	18	4	–	J-POWER	1973	
9	Ninokura	Gonohe	Aomori	FA	C	37	106	186	20	3	21	Pref. Gov.	1970	
10	Kou-noyama	Shinano	Niigata	FA	H	33	380	325	7	0.6	1	TEPCO	1971	

* Reference Number corresponds with the number in the table "Main Dimension of Dams"
Based on the data of Dam Almanac 2007, Japan Dam Foundation.

Outline of Water Resources Development River Systems and facilities

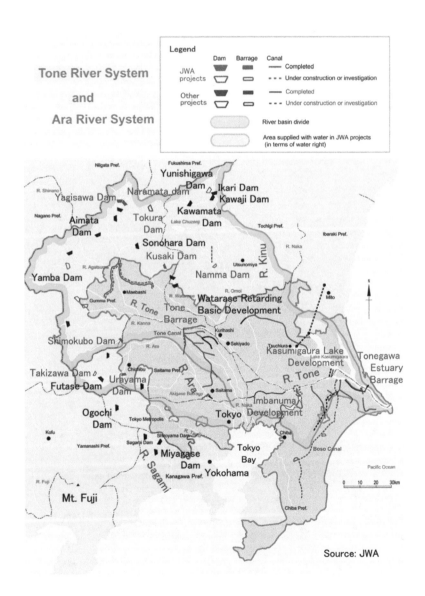

Source: JWA

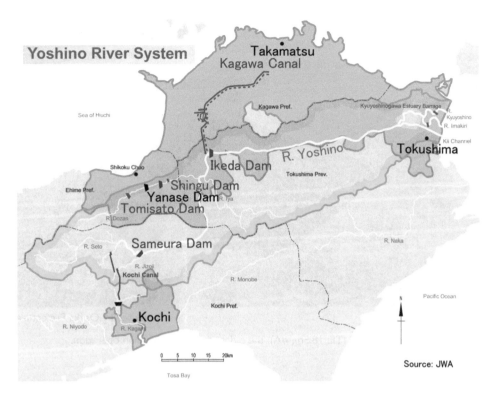

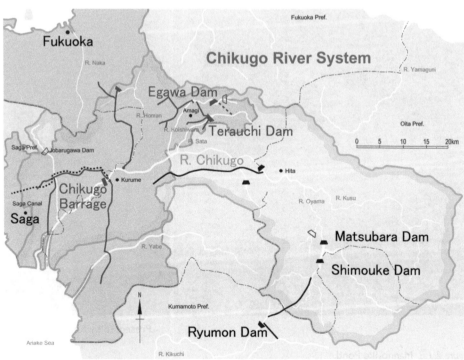

COLOUR PLATES

Presented by Osaka Prefecture

Figure 1.1.1 Sayama-ike Dam (TE, 18.5m, #A): the oldest dam in Japan still in operation.

Presented by Kagawa Prefecture

Figure 2.1.3 Manno-ike Pond.

Presented by MAFF

Figure 2.1.4 Honen-ike Dam.

Presented by MAFF

Figure 2.1.5 Sannokai Dam.

Presented by TEPCO

Figure 2.3.3 Ono Dam of the Yatsuzawa Power Plant.

Presented by CEPCO

Figure 2.3.5 Yasuoka Dam.

Presented by MLIT

Figure 2.4.2 Ikari Dam.

Presented by CEPCO

Figure 2.4.3 Ikawa Dam.

Presented by MLIT

Figure 2.4.5 Ishibuchi Dam.

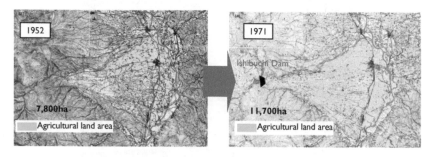

Figure 2.4.6 Change of the agricultural land area on the Isawa Fan.

Source: MLIT.

Presented by MLIT

Figure 2.4.7 Gosho Dam.

Presented by J-POWER

Figure 2.4.8 Sakuma Dam.

Presented by JWA

Figure 2.4.11 Makio Dam.

Presented by JWA

Figure 2.5.4 Yagisawa Dam.

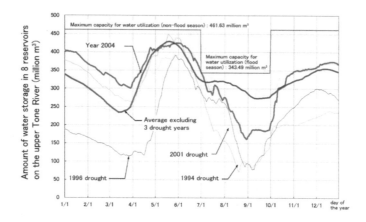

Figure 2.5.7 Replenishment by dams on the Upstream Tone River in 2004.

Source: MLIT.

Figure 2.5.8 Seta River Weir.

Source: MLIT.

Presented by J-POWER

Figure 2.5.12 Okutadami Dam.

Presented by KEPCO

Figure 2.5.13 Kurobe Dam.

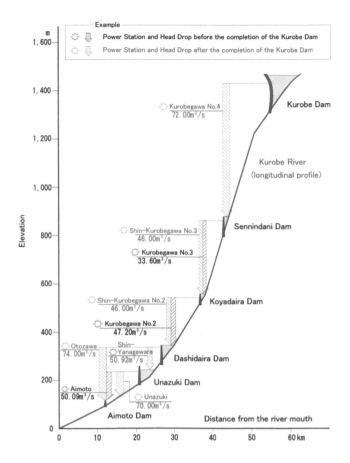

Figure 2.5.14 Schematic view of hydropower generation on the Kurobe River.

Source: KEPCO.

Presented by TEPCO

Figure 2.5.16 Takase Dam.

Presented by JWA

Figure 2.5.19 Sameura Dam.

Presented by KEPCO

Figure 3.3.1 Sho River Unified Intake Weir.

Presented by KEPCO

Figure 3.3.7 Sand flushing at the Dashidaira Dam.

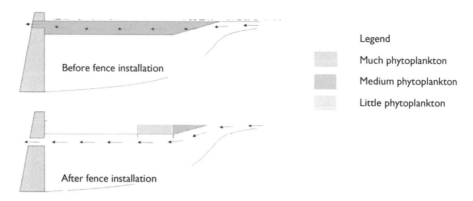

Legend

Much phytoplankton

Medium phytoplankton

Little phytoplankton

Figure 3.3.11 Function of reservoir partition fence (with and without).

Source: JWA.

Presented by JWA

Figure 3.3.12 Installation of reservoir partition fence.

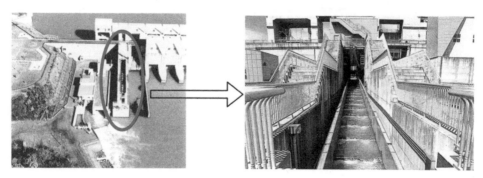

Presented by MLIT

Figure 3.4.3 Fishway of Nibutani Dam.

Figure 3.4.4 Mangrove forests downstream from the Kanna Dam.

Figure 3.5.1 Nagaragawa Estuary Barrage.

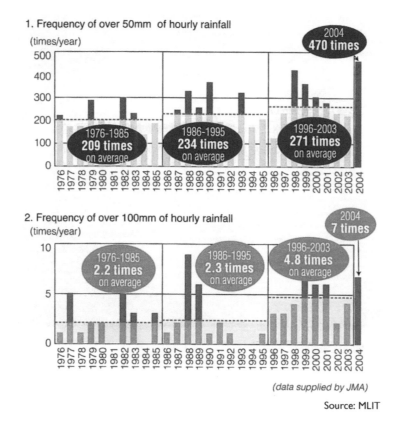

1. Frequency of over 50mm of hourly rainfall

(times/year)

1976-1985
209 times
on average

1986-1995
234 times
on average

1996-2003
271 times
on average

2004
470 times

2. Frequency of over 100mm of hourly rainfall
(times/year)

1976-1985
2.2 times
on average

1986-1995
2.3 times
on average

1996-2003
4.8 times
on average

2004
7 times

(data supplied by JMA)

Source: MLIT

Figure 4.1.3 Frequency of concentrated rainfall.

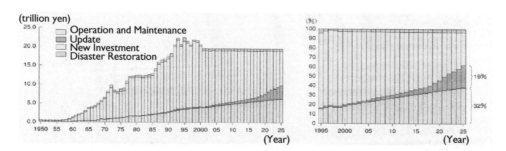

(trillion yen)

Operation and Maintenance
Update
New Investment
Disaster Restoration

(Year)

(Year)

Figure 4.1.5 Estimation of investment demand for maintenance and renewal of public infrastructure.
Source: MLIT.

Figure 4.3.1 Hakusui Dam.

Presented by Mr. Masahisa Okano

Figure 4.4.1 Proserpina Dam

Location of dams in Japan

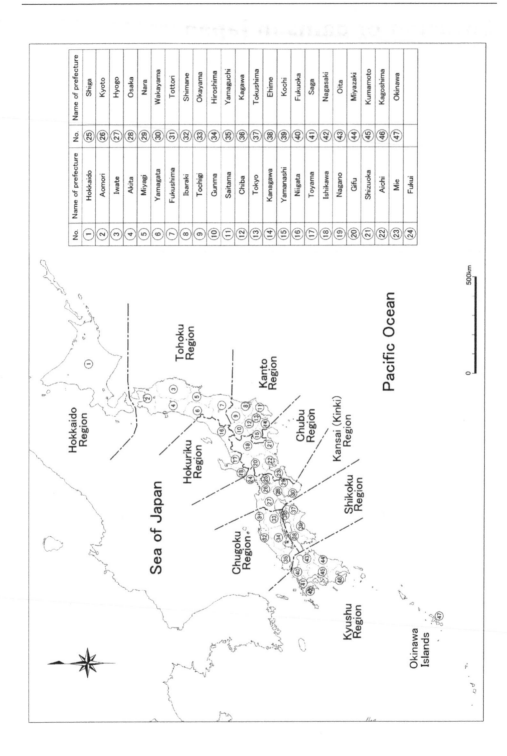

No.	Name of prefecture	No.	Name of prefecture
①	Hokkaido	㉕	Shiga
②	Aomori	㉖	Kyoto
③	Iwate	㉗	Hyogo
④	Akita	㉘	Osaka
⑤	Miyagi	㉙	Nara
⑥	Yamagata	㉚	Wakayama
⑦	Fukushima	㉛	Tottori
⑧	Ibaraki	㉜	Shimane
⑨	Tochigi	㉝	Okayama
⑩	Gunma	㉞	Hiroshima
⑪	Saitama	㉟	Yamaguchi
⑫	Chiba	㊱	Kagawa
⑬	Tokyo	㊲	Tokushima
⑭	Kanagawa	㊳	Ehime
⑮	Yamanashi	㊴	Kochi
⑯	Niigata	㊵	Fukuoka
⑰	Toyama	㊶	Saga
⑱	Ishikawa	㊷	Nagasaki
⑲	Nagano	㊸	Oita
⑳	Gifu	㊹	Miyazaki
㉑	Shizuoka	㊺	Kumamoto
㉒	Aichi	㊻	Kagoshima
㉓	Mie	㊼	Okinawa
㉔	Fukui		

Main dimensions of dams

No.	Name of dams	Dam type	Purpose	Catchment area (km²)	Dam height (m)	Storage capacity (10³ m³)	Owner	Fiscal year of completion
1	Nokanan	PG	H	113.3	45.5	15,300	HEPCO	1971
2	Jozankei	PG	CIH	104.0	117.5	82,300	MLIT	1989
3	Hoheikyo	PG	CIH	159.0	102.5	47,100	MLIT	1972
4	Izarigawa	ER	CNS	113.3	45.5	15,300	MLIT	1980
5	Pirika	PG/ER	CNIH	115.0	40.0	18,000	MLIT	1991
6	Nibutani	PG	CNSIH	1,215.0	32.0	27,100	MLIT	1997
7	Samani	PG	C	54.9	44.0	6,200	Pref. Gov.	1974
8	Sasanagare	CB	S	–	25.3	607	Pref. Gov.	1923
9	Aseishigawa	PG	CNSH	225.5	91.0	53,100	MLIT	1988
10	Okiura	PG	CNH	200.8	40.0	3,583	Pref. Gov.	1944
11	Shijushida	PG/ER	CH	1,196.0	50.0	47,100	MLIT	1968
12	Gosho	PG/ER	CNSH	635.0	52.5	65,000	MLIT	1981
13	Sannokai	ER	I	37.7	61.5	38,400	MAFF	2001
14	Yuda	VA	CNH	583.0	89.5	114,160	MLIT	1964
15	Tase	PG	CNH	740.0	81.5	146,500	MLIT	1954
	Tono	PG	C	29.6	26.5	1,225	Pref. Gov.	1957
16	Ishibuchi	CFRD	CNH	154.0	53.0	16,150	MLIT	1953
17	Kamafusa	PG	CNSIH	195.3	45.5	45,300	MLIT	1970
18	Gassan	PG	CNSH	239.8	123.0	65,000	MLIT	2001
19	Sagae	ER	CNSIH	231.0	112.0	109,000	MLIT	1990
20	Shirakawa	ER	CNSIH	205.0	66.0	50,000	MLIT	1981
21	Miharu	PG	CNSIH	226.4	65.0	42,800	MLIT	1997
22	Okawa	PG	CNSIH	825.6	75.0	57,500	MLIT	1987
23	Mikawasawa	PG	CN	–	48.5	899	Pref. Gov.	2003
	Yunishigawa	PG	CNSI	102.0	130.0	99,000	MLIT	2011

(Continued)

(Continued)

No.	Name of dams	Dam type	Purpose	Catchment area (km²)	Dam height (m)	Storage capacity (10³ m³)	Owner	Fiscal year of completion
24	Kawaji	VA	CNSIH	323.6	140.0	83,000	MLIT	1983
25	Ikari	PG	CNH	271.2	112.0	55,000	MLIT	1956
26	Kurobe (Tochigi)	PG	H	267.3	33.9	2,366	TEPCO	1912
27	Kawamata	VA	CNH	179.4	117.0	87,600	MLIT	1966
28	Nakaiwa	VA	H	697.0	26.3	1,488	TEPCO	1924
29	Namma	CFRD	CNSI	344.7	86.5	51,000	JWA	2010
30	Watarase Reservoir	–	CNS	8,588.0	–	26,400	MLIT	2002
31	Yagisawa	VA	CHIS	167.4	131.0	204,300	JWA	1967
32	Naramata	ER	CHIS	95.4	158.0	90,000	JWA	1991
33	Tokura	PG	–	–	–	–	JWA	–
34	Fujiwara	PG	CNH	401.0	95.0	52,490	MLIT	1958
35	Kusaki	PG	CHIS	254.0	140.0	60,500	JWA	1977
36	Sonohara	PG	CNH	607.6	76.5	20,310	MLIT	1965
37	Tambara	ER	H	6.5	116.0	14,800	TEPCO	1982
38	Aimata	PG	CNH	110.8	67.0	25,000	MLIT	1959
39	Yamba	PG	CS	707.9	131.0	107,500	MLIT	2010
40	Shimokubo	PG	CHS	322.9	129.0	130,000	JWA	1968
41	Takizawa	PG	CHS	108.6	140.0	63,000	JWA	2008
42	Futase	VA	CNH	170.0	95.0	26,900	MLIT	1961
43	Urayama	PG	CHS	51.6	156.0	58,000	JWA	1999
44	Yamaguchi	TE	S	7.2	35.0	20,649	Pref. Gov.	1934
45	Tonegawa Estuary Barrage	Weir	CIS	–	–	–	JWA	1971
46	Imbanuma Development	–	IS	–	–	–	JWA	1968
47	Murayama Reservoir	TE	S	1.3	24.2	3,321	Pref. Gov.	1924
		TE	S	2.0	32.6	12,148		1927
48	Shiromaru	PG	H	397.0	30.3	893	Pref. Gov.	1963
49	Ogochi	PG	SH	262.9	149.0	189,100	Pref. Gov.	1957
50	Shiroyama	PG	CSH	1,221.3	75.0	62,300	Pref. Gov.	1965
51	Sagami	PG	SH	1,128.5	58.4	63,200	Pref. Gov.	1947
52	Miyagase	PG	CNSH	213.9	156.0	193,000	MLIT	2001

(Continued)

(Continued)

No.	Name of dams	Dam type	Purpose	Catchment area (km²)	Dam height (m)	Storage capacity (10³ m³)	Owner	Fiscal year of completion
53	Myoken Weir	–	NSH	–	–	–	MLIT	1990
54	Unazuki	PG	CSH	617.5	97.0	24,700	MLIT	2001
55	Dashidaira	PG	H	461.2	76.7	9,010	KEPCO	1985
56	Koyadaira	PG	H	404.80	51.5	2,122	KEPCO	1936
57	Sennindani	PG	H	284.09	43.5	682	KEPCO	1940
58	Kurobe (Toyama)	VA	H	184.5	186.0	199,285	KEPCO	1963
59	Komaki	PG	H	1,100.0	79.2	37,957	KEPCO	1930
60	Tedorigawa	ER	CSH	428.2	153.0	231,000	J-POWER	1979
61	Managawa	VA	CNH	223.7	127.5	115,000	MLIT	1977
62	Ono	TE	H	5.9+582.0*	37.3	1,692	TEPCO	1914
63	Takase	ER	H	131.0	176.0	76,200	TEPCO	1979
64	Misogawa	ER	CHS	55.1	140.0	61,000	JWA	1996
65	Miwa	PG	CSH	311.1	69.1	34,300	MLIT	1959
66	Makio	ER	HIS	304.0	105.0	75,000	JWA	1961
67	Yasuoka	PG	H	2,980.0	50.0	10,761	CEPCO	1936
68	Hosobidani	PG	H	1.8	22.4	71	CEPCO	1926
69	Agigawa	ER	CS	82.0	101.5	48,000	JWA	1991
70	Oi	PG	H	2,083.0	53.4	29,400	KEPCO	1924
71	Maruyama	PG	CH	2,409.0	98.2	79,520	KEPCO	1955
72	Imawatari	PG	H	4,632.3	34.3	9,470	KEPCO	1939
73	Kaore	VA	H	2.5	107.5	17,200	CEPCO	1995
74	Kamiosu	ER	H	12.0	98.0	14,500	CEPCO	1995
75	Tokuyama	ER	CHS	254.5	161.0	660,000	JWA	2008
76	Tashiro	PG	H	108.0	17.3	220	TEPCO	1928
77	Ikawa	HG	H	459.3	103.6	15,000	CEPCO	1957
78	Nagashima	PG	CNIS	534.3	109.0	78,000	MLIT	2002
79	Oigawa	PG	H	537.0	33.5	788	CEPCO	1936
80	Sumatagawa	PG	H	240.9	34.8	987	CEPCO	1936
81	Sasamagawa	PG	H	68.0	46.4	6,340	CEPCO	1960
82	Sakuma	PG	H	4,156.5	155.5	326,850	J-POWER	1956
83	Akiha	PG	HIS	4,490.0	89.0	34,700	J-POWER	1958
84	Yahagi	VA	CNSIH	504.5	100.0	80,000	MLIT	1970
85	Shitara	PG	CINS	62.2	129.0	98,000	MLIT	2020

(Continued)

(*Continued*)

No.	Name of dams	Dam type	Purpose	Catchment area (km²)	Dam height (m)	Storage capacity (10³ m³)	Owner	Fiscal year of completion
86	Nagaragawa Estuary Barrage	Weir	CS	–	–	–	JWA	1995
87	Hinachi	PG	CHS	75.5	70.5	20,800	JWA	1999
	Hachisu	PG	CNSH	80.9	78.0	32,600	MLIT	1991
88	Lake Biwa Project	–	CS	3,848.0	–	–	JWA	1992
89	Hiyoshi	PG	CS	290.0	70.4	66,000	JWA	1998
90	Kurokawa	ER	H	5.2	98.0	26,151	KEPCO	1974
91	Hitokura	PG	CS	115.1	75.0	33,300	JWA	1983
92	Sengari	PG	S	94.5	42.4	11,610	Pref. Gov.	1919
93	Aono	PG	CNS	51.8	29.0	15,100	Pref. Gov.	1987
94	Gohonmatsu	PG	S	10.7	33.3	417	Pref. Gov.	1900
95	Tachigahata	PG	S	19.8	33.3	1,248	Pref. Gov.	1905
96	Nunome	PG	CS	75.0	72.0	17,300	JWA	1992
97	Murou	PG	CS	136 + 33*	63.5	16,900	JWA	1974
98	Otaki	PG	CNSH	258.0	100.0	84,000	MLIT	2009
99	Sarutani	PG	NH	214.9	74.0	23,300	MLIT	1957
100	Asahi	VA	H	39.2	86.1	16,920	KEPCO	1978
101	Tomata	PG	CNSH	217.4	74.0	84,100	MLIT	2004
102	Shin-Nari-wagawa	PG	HS	625.0	103.0	127,500	ENERGIA	1968
103	Taishakugawa	PG	H	120.0	62.4	14,278	ENERGIA	1931
104	Kodo	PG	HCS	152.0	43.0	7,030	Pref. Gov.	1940
105	Shimajigawa	PG	CNSI	32.0	89.0	20,600	MLIT	1981
106	Kotogawa	PG	CHIS	324.0	38.8	23,788	Pref. Gov.	1950
107	Kyu-Yoshinoga-wa Estuary Barrage	Weir	C	–	–	–	JWA	1976
108	Ikeda	PG	CH	1,904.0	24.0	12,650	JWA	1975
109	Nagayasuguchi	PG	CNH	582.9	85.5	54,278	Pref. Gov.	1956
110	Honen-ike	CB	CI	8.0	30.4	1,643	Pref. Gov.	1930
111	Shingu	PG	CHIS	215 + 39*	42.0	13,000	JWA	1975
112	Yanase	PG	CSIH	170.7	55.5	46,260	Pref. Gov.	1953
113	Tomisato	PG	CSH	101.2	111.0	52,000	JWA	2000
114	Ishitegawa	PG	CSI	72.6	87.0	12,800	MLIT	1972

(*Continued*)

(Continued)

No.	Name of dams	Dam type	Purpose	Catchment area (km²)	Dam height (m)	Storage capacity (10³ m³)	Owner	Fiscal year of completion
115	Yamatosaka	PG	CN	64.7	103.0	24,900	MLIT	2019
116	Nomura	PG	CIS	168.0	60.0	16,000	MLIT	1981
117	Sameura	PG	CNSIH	527.0	106.0	316,000	JWA	1977
118	Yanase	ER	H	100.7	115.0	104,625	J-POWER	1970
119	Terauchi	ER	CIS	51.0	83.0	18,000	JWA	1978
120	Meboro	PG	CN	14.2	40.0	5,400	Pref. Gov.	1999
121	Kayaze	PG	CS	18.9	51.0	3,030	Pref. Gov.	1961
122	Ogakura	PG	CNS	4.3	41.2	2,040	Pref. Gov.	1926
123	Yukinoura	PG	CNS	19.9	44.0	3,900	Pref. Gov.	1976
124	Konoura	PG	CIS	16.5 + 8.5*	51.0	6,840	Pref. Gov.	1970
125	Hongochi Upper	ER	S	3.5	28.2	496	Pref. Gov.	1891
	Hongochi Lower	PG	S	4.6	26.9	616	Pref. Gov.	1904
126	Nishiyama	PG	S	–	40.0	1,580	Pref.Gov.	1905
127	Matsubara	PG	CNSH	491.0	83.0	54,600	MLIT	1986
128	Shimouke	VA	CNH	185.0	98.0	59,300	MLIT	1958
129	Kamishiiba	VA	H	279.6	110.0	91,550	Kyushu EPCO	1955
130	Matsuo	PG	CNH	304.1	68.0	45,202	Pref. Gov.	1951
131	Kawabegawa	VA	CNIH	470.0	107.5	133,000	MLIT	–
132	Setoishi	PG	H	1,629.3	26.5	9,930	J-POWER	1958
133	Tsuruta	PG	CH	805.0	117.5	123,000	MLIT	1959
134	Taiho	PG	CNS	13.3	77.5	20,050	Cabinet Office	2009
135	Fukuji	ER	CNS	32.0	91.7	55,000	Cabinet Office	1990
136	Kanna	PG	CNIS	7.6	45.0	8,200	Cabinet Office	1993
A	Sayama-ike Pond	TE	I	17.9	18.5	2,800	Pref. Gov.	616
	Sayama-ike	TE	CN	17.9	18.5	2,800	Pref. Gov.	2001
B	Futakawa	PG	CH	228.8	67.4	30,100	Pref. Gov.	1966

Type of dam   **ER**: Rockfill **TE**: Earthfill **PG**: Gravity **VA**: Arch **CB**: Buttress **HG**: Hollow Gravity

Purpose	**C:** Flood control **H:** Hydroelectric **I:** Irrigation **S:** Water supply
	N: Normar functions of the river water
Owner	**MLIT:** Ministry of Land, Infrastructure, Transport and Tourism
	MAFF: Ministry of Agriculture Forestry and Fisheries
	JWA: Japan Water Agency
	Pref.Gov.: Prefectural Government
	HEPCO: Hokkaido Electric Power Co., Inc.
	TEPCO: Tokyo Electric Power Co., Inc.
	CEPCO: Chubu Electric Power Co., Inc.
	Hokuriku EPCO: Hokuriku Electric Power Co., Inc.
	KEPCO: Kansai Electric Power Co., Inc.
	ENERGIA: Chugoku Electric Power Co., Inc
	YONDEN: Shikoku Electric Power Co., Inc
	Kyushu EPCO: Kyushu Electric Power Co., Inc
	J-POWER: Electric Power Deveropment Co., Ltd.

* the former figure: direct catchment area, the latter figure: indirect catchment area

References

[1] *Chronology of Japanese History.* Kodansha Co., Ltd., Tokyo, 2005.

[2] Takahashi,Y.: *History of Modern Japanese Civil Engineering* (in Japanese). Shokokusha Publishing Co., Ltd., Tokyo, 2004.

[3] Goda, Y.: *Civil Engineering and Civilization* (in Japanese). Kajima Institute Publishing Co., Ltd., Tokyo, 1996.

[4] Takebayashi, S.: History of Dam Engineering (in Japanese). In: Japan Association of Dam & Weir Equipment Engineering (eds): Compendium of Gates, Vol. I, Book of Description, Japan Association of Dam & Weir Equipment Engineering, 1987.

[5] World Commission on Dams: *DAMS AND DEVELOPMENT – A New Framework for Decision-making*, THE REPORT OF THE WORLD COMMISSION ON DAMS. Earthscan Publications Ltd., 2000.

[6] River Bureau, MLIT (2005). Comparisons of Reservoir Capacity between Japan and U.S. [Online]. Available: http://www.mlit.go.jp/river/dam/main/opinion/america_dam/graph01.html

[7] WRDPC: *With Water – 40 Years Footprint of Water Resources Development Public Corporation and Leap toward New Century* (in Japanese), 2003.

[8] Amano, R.: *Dams and Japan* (in Japanese). Iwanami Shoten, Publishers, Tokyo, 2001.

[9] *Citizens' Guide to the World Commission on Dams* (in Japanese). RWESA-Japan, 2002.

[10] Editorial Committee for Meteorology of Japan: *Rika Nenpyo (Chronological Scientific Tables) – Meteorology of Japan (CD-ROM)*. Maruzen Co. Ltd., Tokyo, 2002.

[11] Water Resources Department, Land and Water Bureau, MLIT: *Water Resources in Japan, 2004 Edition* (in Japanese), 2004.

[12] River Bureau, MLIT (2003). Dam Circumstances of the United States [Online]. Available: http://www.mlit.go.jp/river/dam/main/opinion/america_dam/america_dam_index.html

[13] Japan Dam Foundation: *Dam Almanac 2001* (in Japanese), 2001.

[14] Shimura, H.: *Irrigation and Land in Japan* (in Japanese). University of Tokyo Press, Tokyo, 1987.

[15] Imamura, N.: *100 Years History of Land Improvement* (in Japanese). Heibonsha Ltd, Publishers, Tokyo, 1977.

[16] The Japanese Society of Irrigation, Drainage and Rural Engineering: *Handbook on Land Drainage and Irrigation, 6th Edition* (in Japanese), 2000.

[17] Sato, M.: History of Irrigation Water in 20th Century (in Japanese). *Kasen (River)*, December 2000 issue, Japan River Association, pp. 64–70.

[18] Sayama-ike Pond Museum, Osaka Prefecture: *Exhibition Record 1*, Permanent Exhibition Guide (in Japanese), 2002.

[19]   Editorial Committee for History of Irrigation Pond in Sanuki Region: *History of Irrigation Pond in Sanuki Region* (in Japanese). GYOSEI Corporation, Tokyo, 2000.

[20]   Yukawa, K.: Development of Fill Type Dam Construction (in Japanese). *Journal of the Japanese Society of Irrigation, Drainage and Reclamation Engineering* (JSIDRE) 49(9), p. 8.

[21]   Hatate, I.: *History of Water Management in Japan* (in Japanese). Association of Agriculture and Forestry Statistics, 1983.

[22]   Taniyama, S.: Technological Development of Irrigation Dam (in Japanese). In: Japan Association of Dam & Weir Equipment Engineering (eds): Compendium of Gates, Vol. I, Book of Description, Japan Association of Dam & Weir Equipment Engineering, 1987.

[23]   Japan Dam Foundation: *Dam Almanac 2003* (in Japanese), 2003.

[24]   Agricultural Structure Improvement Bureau, MAFF: *Survey on Long-ranged Disaster Prevention Projects for Irrigation Pond* (in Japanese), 1996.

[25]   Editing Office, Shikoku Shimbun, et al.: *Irrigation Pond in Sanuki Region* (in Japanese). Bikohsha, Takamatsu, 1975.

[26]   Manno-ike Land Improvement District: *History of Manno-ike Pond* (in Japanese), 2001.

[27]   Manno-ike Land Improvement District (2002). Manno-ike [Online]. Available: //www9. ocn.ne.jp/~mannoike/index.html

[28]   Sannokai National Irrigation Project Office, Tohoku Regional Agricultural Administration Office: *Water of Kuzumaru and Sannokai* (in Japanese), 2002.

[29]   The Japanese Society of Irrigation, Drainage and Rural Engineering: *Record of Agricultural Water Use in the Kitakami River System* (in Japanese), 1995.

[30]   Sannokai National Irrigation Project Office, Tohoku Regional Agricultural Administration Office: *Record of Sannokai Dam Engineering* (in Japanese), 2002.

[31]   Japan Water Works Association: *Outline of Waterworks 2001* (in Japanese), 2001.

[32]   Ministry of Welfare: *100 Years History of Waterworks* (in Japanese), 1990.

[33]   ditto 31, pp. 3–4.

[34]   Japan Water Works Association: *History of Waterworks in Japan*, Book of Particular Area 3 (in Japanese), 1967.

[35]   Waterworks Bureau, Nagasaki City (2004). History of Waterworks in Nagasaki [Online]. Available: http://www1.city.nagasaki.nagasaki.jp/water/mizu/index.html

[36]   Waterworks Bureau, Nagasaki City: *100 Years History of Waterworks in Nagasaki* (in Japanese), 1992.

[37]   Waterworks Bureau, City of Kobe: *100 Years History of Waterworks in Kobe* (in Japanese), 2001.

[38]   Waterworks Bureau, City of Kobe (2004). 100 Years Footsteps in Kobe Waterworks [Online]. Available: http://www.city.kobe.jp/cityoffice/51/06/09.html

[39]   ditto 31, p. 25.

[40]   ditto 31, p. 24.

[41]   ditto 31, p. 37.

[42]   River Bureau, MLIT (2003). About Dam Projects [Online]. Available: http://www.mlit. go.jp/river/dam/main/dam/ref-e2.html

[43]   Ministry of Welfare: *100 Years History of Waterworks* (in Japanese), 1990.

[44]   Bureau of Waterworks, Tokyo Metropolitan Government: *100 Years History of Modern Waterworks in Tokyo*, History of Particular Division (in Japanese), 1999.

[45]   Japan Water Works Association: *History of water works in Japan*, Book of particular 1 (in Japanese), 1967.

[46]   Bureau of Waterworks, Tokyo Metropolitan Government: *100 Years History of Modern Waterworks in Tokyo*, History of Particular Division (in Japanese), 1999.

[47]   Japan Dam Foundation: *Dam Almanac 2005* (in Japanese), 2005.

[48]   Bureau of Waterworks, Tokyo Metropolitan Government: *100 Years History of Modern Waterworks in Tokyo*, Data and Chronology (in Japanese), 1999.

[49]   Ida, T.: The effect of reservoir development in Sapporo City (in Japanese). *Engineering for Dams*, No. 202, pp. 72–80.

[50]   Japan Electric Power Civil Engineering Association: *100 Years History of Hydropower Technology* (in Japanese), 1992.

[51]   ditto 50, pp. 29, 39.

[52]   ditto 50, pp. 1065–1137.

[53]   ditto 50, pp. 15, 28, 39.

[54]   ditto 50, p. 25.

[55]   Kosaka, T.: *Research on Modern River Improvement of the Tone River from the View Point of Planning Methodology* (in Japanese). Doctoral Thesis of the University of Tokyo, p. 133.

[56]   Mononobe, N.: Characteristics of Water Retention Dam and its Rational Design Method (in Japanese). *Civil Engineering*, JSCE, Vol. 11, No. 5.

[57]   Japan Dam Engineering Center: *Construction of Multi-Purpose Dam*, Vol. 1, Book of Planning and Administration (1987 version) (in Japanese), 1987.

[58]   MLIT: *Program Evaluation of Dam Projects – Validation of Effects and Impacts on the Region –* (in Japanese), 2003.

[59]   ditto 50, pp. 57–58.

[60]   ditto 50, pp. 1065–1137.

[61]   Takasaki, T.: Tower of Theory and Tower of Technology (in Japanese). *The Dam Dijest*, No. 730, pp. 7–9.

[62]   Japan Dam Foundation: *Dam Almanac 2005* (in Japanese), 2005.

[63]   Tsumori, T.: Effects of Constructing Reservoirs in the Kitakami River Basin (in Japanese). Proceeding of the Session "The Effects of Providing Reservoir for River Basin Development", 3rd World Water Forum, Kyoto, pp. 31–45.

[64]   J-POWER: *30 Years History of J-POWER* (in Japanese), 1984.

[65]   Hasebe, N.: *Sakuma Dam – Historical Record* (in Japanese). Toyo Shokan, Tokyo, 1956.

[66]   Japan Human Science Society: *Sakuma Dam* (in Japanese). University of Tokyo Press, Tokyo, 1958.

[67]   Hamamatsu Work Office, MOC: *Tenryu River, Flood Control and Water Utilization* (in Japanese), 1990.

[68]   Kubota, M.: *With the Tenryu River – Geography, Geology and Those Who Challenged its Torrent* (in Japanese). Chunichi Shimbun Co. Ltd., Nagoya, 2001.

[69]   CEPCO: *Oi River – Culture in its Basin and Electricity* (in Japanese), 2001.

[70]   Aichi Canal Dept., WRDPC: *30th Anniversary of Aichi Canal Project – Appreciation for the Ample Water* (in Japanese), 1991.

[71]   Investigation Office for Agricultural Water Use in the Kiso River System, Tokai Regional Agricultural Administration Office, MAFF: *Record of Agricultural Water Use in the Kiso River System* (in Japanese), 1980.

[72]   Aichi Irrigation Public Corporation: *History of the Aichi Canal Project* (in Japanese), 1968.

[73]   ditto 71.

[74]   ditto 71, p. 336.

[75]   Chubu Regional Bureau, WRDPC: *Outline of Water Use Plan of the Aichi Canal Second Phase Project* (in Japanese), 1982.

[76]   Yamamoto Saburo: *Historical Study on Modern River Project leading to the Overall Revision of the River Law* (in Japanese), 1993.

[77]   ditto 76, p. 268.
[78]   Water Resources Department, Land and Water Bureau, MLIT: *Water Resources in Japan, 2005 Edition* (in Japanese), 2005.
[79]   JWA: *Outline of Works in 2004* (in Japanese), 2004.
[80]   ditto 76, p. 266.
[81]   Nakazawa, K., et al.: *Case Study of Water Resources* (in Japanese). System of Civil Engineering Vol. 24, Shokokusha Publishing Co., Ltd, Tokyo, 1978, pp. 219–236.
[82]   Ikeda Comprehensive Administration Office, WRDPC: *Construction Record of the Sameura Dam* (in Japanese), 1979.
[83]   River Bureau, MLIT (2004). Necessity and Effect of Dam [Online]. Available: //www. mlit.go.jp/river/gaiyou/panf/index.html
[84]   Gosho Dam Construction Office, Tohoku Regional Construction Bureau, MOC: *Construction Record of the Gosho Dam* (in Japanese), 1982.
[85]   Agency for Natural Resources and Energy, MITI: *Outline of Electricity Demand*, Fiscal 1951 to 1965 (in Japanese), Fiscal 1951 to 1965.
[86]   The Electric Power Civil Engineering Association: *100 Years History of Hydropower Technology* (in Japanese), 1992.
[87]   ditto 86, pp. 1065–1137.
[88]   ditto 86, pp. 94–139.
[89]   KEPCO: *100 Years History of Hydropower Technology of KEPCO* (in Japanese).
[90]   ditto 86, pp. 76–81.
[91]   Agency for Natural Resources and Energy, METI: *Outline of Electricity Development, Fiscal 2002* (in Japanese), 2002.
[92]   FEPC (2004.3). Best Mix of Electric Power Sources [Online]. Available: //www.tepco. co.jp/custom/LapLearn/ency/cmb01-j.html2003.9,//www.tepco.co.jp/
[93]   ditto 83.
[94]   Japan River Association: *River Handbook 2003* (in Japanese), 2003.
[95]   Japan Dam Foundation: *Dam Almanac 2005* (in Japanese), 2005.
[96]   MLIT: *Pogram Evaluation Report of Dam Projects – Inspection of Effects Acting on the Regions* (in Japanese), 2003.3.
[97]   River Department, Shikoku Regional Development Bureau, MLIT: *Floods in the Shikoku Region by the Typhoon Chaba, No. 16, of 2004* (in Japanese), 2004.9.
[98]   WEC: *Management of Dam – Compilation of Regulations 2003* (in Japanese). Sankaido Publishing Co., Ltd., Tokyo, 2003.
[99]   Committee of Comprehensive Policies against Downpour Disasters, Subcouncil of Rivers, Council of Social Infrastructure Development (2005.4). Proposal for Comprehensive Measures against Downpour Disasters [Online]. Available: //www.m
[100]   WEC: About Measures for Reservoir Area Development (in Japanese), 1996.
[101]   Study Group on Reservoir Area: *Reservoir Area Vision in 21st Century* (in Japanese). WEC, 1999.
[102]   ditto 101, pp. 41–77.
[103]   Gosho Dam Construction Office, Tohoku Regional Construction Bureau, MOC: *Construction Record of the Gosho Dam* (in Japanese), 1982.
[104]   Sagami River System Dam Management Office, Kanto Regional Development Bureau, MLIT (2004). Miyagase Dam [Online]. Available: http://www.ktr.mlit.go.jp/sagami/ dam/base/index.htm
[105]   Okayama Prefecture (2005). Outline of the Tomata Dam [Online]. Available: http:// www.pref.okayama.jp/soshiki/detail.html?lif_id=4984
[106]   Kawabegawa Dam Construction Office, Kyushu Regional Development Bureau, MLIT (2001). Progress report of the project [Online]. Available: http://www.qsr.mlit.go.jp/ kawabe/qa1-2.html

[107] ditto 101, pp. 72–77.
[108] ditto 101, p. 69.
[109] Maruyama, T.: *Compensation of Dam Project and Reservoir Area Development Plan* (in Japanese). Japan Dam Foundation, 1986.
[110] ditto 101, pp. 118–119.
[111] ditto 101, pp. 83–85.
[112] ditto 109, p. 213.
[113] WEC: *Management of Dam – Compilation of Regulations 2003* (in Japanese). Sankaido Publishing Co., Ltd., Tokyo, 2003.
[114] Takayama, S.: Exchange between Reservoir Area and Downstream Area of the Yamba Dam (in Japanese). *Dam and Water Resources Net*, 2004.5, WEC, pp. 12–13.
[115] ditto 113, p. 658.
[116] ditto 113, pp. 659–666.
[117] ditto 114, p. 10.
[118] Study Group on Irrigation Water Use: *Irrigation Water in Japan* (in Japanese), 1980.
[119] Shinzawa, K.: *Cordination Theory of River Water Use* (in Japanese). Iwanami Shoten, Publishers, Tokyo, 1962.
[120] ditto 119, pp. 147–156.
[121] Shinanogawa River Office, Hokuriku Regional Development Bureau, MLIT. Introduction of Facilities – Myoken Weir [Online]. Available: http://www.hrr.mlit.go.jp/shinano/office/sisetu/index.html
[122] Hanayama, Y. and Fuse, T.: *City and Water Resources* (in Japanese). Kajima Institute Publishing Co., Ltd., Tokyo, 1977.
[123] ditto 122, pp. 83–106.
[124] Japan River Association: *Technical Guideline for River Works of MOC, Book of Planning* (in Japanese). Sankaido Publishing Co., Ltd., Tokyo, 1997.
[125] River Environment Division, River Bureau, MLIT: *Guideline for Studying Normal Flow Rate* (in Japanese), 2001.
[126] River Bureau, MLIT: *About River Maintenance Flow at the Downstream of Hydropower Dams* (in Japanese), 2003.
[127] River Bureau, MLIT (2004). Major Challenges concerning Dam [Online]. Available: http://www.mlit.go.jp/river/dam/main/dam/ref16/ref-p2.html
[128] Flood Prevention Union of Kawaji, Iida City, Nagano Pref.: *History of Flood Fighting in Kawaji Area, Tenryu River, Sequel* (in Japanese), 2003.
[129] Working Group for the Sedimentation Measures, Technical Committee, JCOLD: Current State and Challenges of Reservoir Sedimentation Measures (in Japanese). *Large Dams*, No. 176, JCOLD, pp. 17–44.
[130] ditto 129, p. 38.
[131] Kikuchi, K., Muranaga, M. and Itagusu, K.: Situation and Measures of Sedimentation of the Sakuma Dam (in Japanese). *Electric Power Civil Engineering*, No. 291, Japan Electric Power Civil Engineering Association, pp. 41–45.
[132] Okano, M., Kikui, M., Ishida, H. and Sumi, T.: Reservoir Sedimentation Management by Coarse Sediment Replenishment below Dams. *Proceedings the Ninth International Symposium on River Sedimentation*, p. 1074.
[133] Tsuboka, S.: We need to understand "Environment concerning dams" and "Function of dams at the flood" from the consideration of "Continuousness" and "Character in daily life" (in Japanese). *Dam and Water Resources Net*, 2005.1, WEC, pp. 12–14.
[134] Takahashi, Y.: *River Engineering* (in Japanese). University of Tokyo Press, Tokyo, 1990.
[135] The Dam Maintenance Management Working Group, Japan Society of Dam Engineers: The Report about Reservoir Sedimentation and Turbid Water (III) (in Japanese). *Journal of Japan Society of Dam Engineers*, Vol. 12, No. 2 (2002), Japan Societ.

[136] Japan Dam Engineering Center: *Construction of Multi-purpose Dam*, Vol. 2, Book of Investigation (1987 version) (in Japanese). Japan Construction Training Center, 1987.

[137] Kataoka, K., Daitoh, H. and Katoh, M.: An Example of Sand Flash for the Dam of Kansai Electric Power, Inc. (in Japanese). *Electric Power Civil Engineering*, No. 286, Japan Electric Power Civil Engineering Association, pp. 35–39.

[138] Numata, Y.: Intermittent Type of Aeration Facility of the Kamafusa Dam – Circulation of Reservoir Water (in Japanese). *Kasen (River)*, No. 574, Japan River Association, pp. 96–103.

[139] Yoshida, N. and Sekine, H.: Systematized Concept on Destratification Systems to Control Algal Blooming (in Japanese). *Report of Water Resources Environment Research Institute 2000*, Water Resources Environment Technology Center, p. 21.

[140] Niwa, K. and Kuno, M.: Principle of Flow Control System and Example of Facility (in Japanese). *Engineering for Dams*, No.91, Japan Dam Engineering Center, pp. 24–31.

[141] Kunii, K., Yamaki, M. and Kaneko, M.: The Field Experiment to Keep Water Quality at Kamafusa Dam (in Japanese). *Engineering for Dams*, No.66, Japan Dam Engineering Center, pp. 24–31.

[142] Amano, K.: Water Quality Conservation Techniques Applied in Dam Projects (in Japanese). *Kasen (River)*, No. 683, Japan River Association, pp. 64–69.

[143] Shimizu, T., Yazawa, K. and Niwa, K.: Water Quality Conservation Measures in Sakura Reservoir, Miharu Dam (in Japanese). *Engineering for Dams*, Japan Dam Engineering Center, pp. 71–81.

[144] Tohoku Regional Development Bureau, MLIT (2005). Improvement of Water Quality in Dam Reservoir [Online]. Available: http://www.thr.mlit.go.jp/bumon/b00037/k00290/river-hp/kasen/forefront/kankyou/suishitu-2.html

[145] Yoshimura, T., Tajima, Y. and Saito, S., et al.: Investigation Report about Water Quality Management in Terauchi Dam Reservoir (in Japanese). *Engineering for Dams*, No. 114, Japan Dam Engineering Center, pp. 71–81.

[146] Asaeda, T., Nimal Priyantha, D.G., Saitoh, S. and Gotoh, K.: A New Technique for Controlling Algal Blooms in the Withdrawal Zone of Reservoirs Using Vertical Curtains. *Ecological Engineering*, Vol. 7, Elsevier, B.V., pp. 95–104.

[147] Kudo, K.: Improvement Technology of Water Quality, This and That (in Japanese). *Environmental Conservation Engineering*, Vol. 29, No. 10, Society of Environmental Conservation Engineering, pp. 772–776.

[148] Arai, O. and Takasu, S.: Conservation of Water Quality in Dam Reservoirs (in Japanese). *Journal of Japan Society of Dam Engineers*, Vol. 7, No. 2 (1997), Japan Society of Dam Engineers, pp. 90–97.

[149] Iseri, Y., Kawabata, Z., Fujimoto, K. and Ito, M.: Control of Plankton by Ultraviolet Radiation (in Japanese). *Journal of Water and Waste*, Vol. 38, No. 4, The Industrial Water Institute, pp. 31–37.

[150] Ikeda, S.: Considering about Environmental Problems around Dams (in Japanese). *Engineering for dams*, No. 181, Japan Dam Engineering Center, pp. 4–10.

[151] Miyaji, D.: *Story of Sweetfish, "Ayu"* (in Japanese). Iwanami Shoten, Publishers, Tokyo, 1960.

[152] Investigation Group of Estuary Resources of the Kiso-Nagara-Ibi Rivers: *Investigation Report of Estuary Resources of the Kiso-Nagara-Ibi Rivers* (in Japanese), 1967.

[153] Editrial Committee for "50 Years History of Ministry of Construction": *50 Years History of Ministry of Construction* (in Japanese), 1998.

[154] MOE: *Annual Report on the Environment in Japan* 1999–2003 (in Japanese).

[155] Harada, J. and Yasuda, N.: Conservation and Improvement of the Environment in Dam Reservoirs (in Japanese). *Report of Water Resources Environwent Research Institute 2002*, Water Resources Environment Technology Center, pp. 66–67.

[156] Subcommittee on the Survey of Dam and Environment Problem, JCOLD: Environmental conservation measures of Dam Projects (in Japanese). *Large Dams*, No.183, JCOLD, p. 15.

[157] Kobayashi, T.: Conservation of Rare Birds Found while Constructing Dam (2nd) (in Japanese). *Engineering for Dams*, No. 201, Japan Dam Engineering Center, pp. 61–68.

[158] Koike, N. and Saito, G.: Fish Conservation Measures (Some Problems and Solutions of Fishways which are Constructed on Dam) (in Japanese). *Report of Water Resources Environment Research Institute 2001*, Water Resources Environment Tech.

[159] ditto 156, p. 16.

[160] Osugi, T. and Urakami, M.: Methods of Downstream Environmental Restoration by Flexible Dam Operation (in Japanese). *Report of Water Resources Environment Research Institute 2001*, Water Resources Environment Technology Center, pp. 68–7

[161] Kakizaki, T.: Recent Report on Nagaragawa Estuary Barrage (in Japanese). *Kasen (River)*, May 2001 Issue, Japan River Association, pp. 55–56.

[162] WRDPC: *With Water – 40 Years Footprint of Water Resources Development Public Corporation and Leap toward New Century* (in Japanese), 2003.9.

[163] Study Group on Public Works Project and Communication: *Testimony – Nagaragawa Estuary Barrage*. The Sankei Shimbun, Tokyo, 2002.

[164] Committee on Analysis Method of Environmental Impact at the Planning Stage of River Projects: *Basic Idea for Analysis Method of Environmental Impact at the Planning Stage of River Projects* (in Japanese). Water Resources Environment Technology Center.

[165] Japan Dam Engineering Center: *Construction of Multi-Purpose Dam*, Vol. 1, Book of Planning and Administration (2005 version) (in Japanese), 2005.

[166] MLIT: *Program Evaluation of Dam Projects – Validation of Effects and Impacts on the Region –* (in Japanese), 2003.

[167] MLIT: *Program Evaluation of Water Use Coordination to Improve River Environment – Improvement of River Section Dried-up by Water Intake* (in Japanese), 2003.

[168] Japan Dam Engineering Center: *Construction of Multi-Purpose Dam*, Vol. 6, Book of Construction (2005 version) (in Japanese), 2005.

[169] Statistics Bureau, Ministry of Internal Affairs and Communications (2004). Change of the World Population [Online]. Available: http://www.stat.go.jp/data/sekai/02.htm

[170] MAFF (2004). Report on Overseas Food Supply and Demand in 2004 [Online]. Available: http://www.maff.go.jp/j/zyukyu/jki/j_rep/pdf/2004kaigai–rep.pdf

[171] Water Resources Department, Land and Water Bureau, MLIT: *Water Resources in Japan, 2004 Edition* (in Japanese), 2004.

[172] Special Committee about Food Problem: *For the Solution of the Food Problem in the New Millennium* (in Japanese), 2000.

[173] World Water Assessment Programme, UNESCO: *Water for People, Water for Life*. The 1st UN World Water Development Report.

[174] Wakabayashi, K.: Composition Change of the World Population and Continuation of the Chinese One-child Policy (in Japanese). *Gakushikai Magazine* No. 844, Gakushikai.

[175] The Institute of Applied Energy (2004). Energy Consumption in the World [Online]. Available: http://www.iae.or.jp/energyinfo/energydata/ data1003

[176] FEPC (2005). Energy Quantity of Natural Resources in the World [Online]. Available: http://www.fepc.or.jp/now/resource/002.html

[177] Intergovernmental Panel on Climate Change: *IPCC Third Assessment Report – Climate Change 2001*, 2002.

[178] National Institute of Population and Social Security Research: *Estimation of Future Population of Japan* (in Japanese), 2002.

[179] National Institute of Population and Social Security Research: *Estimation of Future Population of Prefectures (from 2000 to 2030)* (in Japanese), 2002.

[180] MAFF (2004). Food Supply and Demand Table in 2003 [Online]. Available: http://www.kanbou.maff.go.jp/www/jikyu/jikyu01.htm

[181] Oki, T.: Water around the Earth, People around the Water. In: Kada, Y. (eds): Man and Nature around the Water (in Japanese). Yuhikaku, Tokyo, 2003.

[182] The Institute of Applied Energy (2004). Energy Balance in Japan [Online]. Available: http://www.iae.or.jp/energyinfo/energydata/data1007

[183] FEPC (2005). Present Conditions of Electric Power Companies [Online]. Available: http://www.fepc.or.jp/thumbnail/zumen/1-22.html

[184] FEPC (2005). Present Conditions of Electric Power Companies [Online]. Available: http://www.fepc.or.jp/thumbnail/zumen/1-20.html

[185] Hosaka, N.: *Illustration of General Knowledge about Abnormal weather* (in Japanese). Natsumesha Co., Ltd., Tokyo, 2000.

[186] River Bureau, MLIT (2005). National Land Structure Vulnerable to Disaster [Online]. Available: http://www.mlit.go.jp/river/gaiyou/yosan/h17budget3/ref1.pdf

[187] Okamura, Y.: New Dam Management with the Improvement of the Function (in Japanese). *Quarterly River Review*, Shin Koronsha, pp. 62–63.

[188] Takebayashi, S.: *Story of Dam, Sequel* (in Japanese). Gihodo Shuppan Co., Ltd., Tokyo, 2004.

[189] Takebayashi, S.: Deplore a View Asserting Uselessness of Dams, part 3(in Japanese). The Nikkan Kensetsu Kogyo Shimbun Co., Ltd., Tokyo, 2005.

[190] MLIT: *White Paper on MLIT in 2003* (in Japanese), 2003.

[191] Atomic Energy Library Web Page (2005). A Change in Electric Energy Generation in Japan [Online]. Available: http://sta-atm.jst.go.jp/atomica/fig_01070508_01.html

[192] Agency for Natural Resources and Energy, METI (2002). Potential Hydropower Capacity in Japan [Online]. Available: http://www.enecho.meti.go.jp/beforeenecho/hydraulic/data/stock/top.html

[193] Water Resources Department, Land and Water Bureau, MLIT: *Water Resources in Japan, 2004 Edition* (in Japanese), 2004.

[194] Development Division, River Bureau, MOC: *Present Stability of River Water Use Supplied by Dam Reservoirs* (in Japanese), 1996.

[195] Commission for Taking Back Public-Works Project to the Public (2000). Opinion: Concept of Green Dam – Creation of 21st Century's Life Style, Coexistence with River (in Japanese) [Online]. Available: http://www.ktroad.ne.jp/~kjc/md.html

[196] Science Council of Japan (2001). Report on the Evaluation of Multiphasic Function of Agriculture and Forest in Relation with the Global Environment and Human Society (in Japanese) [Online]. Available: http://www.scj.go.jp/ja/info/kohyo/pdf/shimon-18-1.pdf

[197] Editorial Board of JSCE: Dream of HATTA Yoichi, which Bloomed at the Jianan Canal (in Japanese). *Civil Engineering*, JSCE, Vol. 88, No. 1, p. 16.

[198] Hamaguchi, T., Nakagawa, H. and Takasu, S.: Important Aspects for Redevelopment of Dams in Japan, Proceedings of 76th Annual Meeting Symposium, 1–46, ICOLD, 2008.

Subject index

NAME OF RIVERS IN JAPAN

NAME OF RIVERS
IN THE WORLD

NAME OF PREFEC-
TURES

T - #0394 - 101024 - C16 - 246/174/14 - PB - 9781138114548 - Gloss Lamination